Rolling Thunder

★ Rolling Thunder ★

BY
MARTIN NORRIS

THE
APPLE
PRESS

A QUINTET BOOK

Published by The Apple Press
6 Blundell Street
London N7 9BH

ISBN 1-85076-370-4

This book was designed and produced by
Quintet Publishing Limited
6 Blundell Street
London N7 9BH

Creative Director: Richard Dewing
Designer: Stuart Walden
Project Editor: Stefanie Foster

Typeset in Great Britain by
Central Southern Typesetters, Eastbourne
Manufactured in Hong Kong by
Regent Publishing Services Limited
Printed in Singapore by Star Standard Industries Pte. Ltd.

The Author would like to thank
The Sussex Coasters; particularly Ben,
Selma and Ray (who, thankfully, does not
know any better at his age).

Contents

Introduction

No question about it, there is a mystique attached to the Harley-Davidson. A simple, air-cooled V-twin engine, based on a design that is over fifty years old, stands out as a complete anachronism in today's multi-cylinder, water-cooled world of motorcycling. Despite this fact, the Harley-Davidson Motor Company is making more models than ever before and they can't sell quick enough.

Not so long ago the company was almost bankrupt and ownership of a Harley-Davidson was considered by many to be antisocial and irresponsible. The riders were frowned upon as rebellious and immature, as having stepped out of society. A whole motorcycle cult developed and the film and media world were quick to make the most of it. Most riders these days, however, portray a very different image while at the same time maintaining the ideal of not quite conforming. Among a gathering of riders you will find characters from every social and economic background, with this fascination in common.

A love of Harley-Davidsons begins the moment you sit astride one and fire it up for the first time. A prod of the starter button and the engine instantly comes to life, the steady beat of the pistons keeps time with your pounding heart and the throaty exhaust note sucks the breath out of your body. A tentative twist of the throttle and the machine responds with an apocalyptic roar. This is an awesome experience never forgotten.

It is now that you realise this is no ordinary motorcycle that you are riding. This is a living beast, and you'll be in that state of euphoria for

ABOVE
This bike started out in 1942 as a WLA, but when the owner was restoring it he incorporated parts from other 45s and brass-plated many of them.

OPPOSITE
Such an image speaks for itself.

days afterwards as you begin to plan just when and how you're going to get your hands on one of your own.

But there is a rumbling in the distance and it is not the sound of a thundering V-twin. Increasingly strict noise and emission restrictions are becoming harder to meet. Every time a little more power is extracted out of the engine in one way, it has to be suffocated in another to meet the tough new laws, and the life and soul of the machine is being slowly squeezed out. Just how much longer the Harley-Davidson will be allowed to exist in its traditional guise remains to be seen.

This book reveals how and why the Harley-Davidson remains the last and greatest American motorcycle ever made. And as for the Harley-Davidson mystique, well, the only way to understand fully what it is, is to own one.

CHAPTER ONE

History

The Harley-Davidson Motor Company was a family affair right from the very beginning, when, in 1903, William (Bill) S. Harley and Arthur Davidson constructed their first motorcycle. They were helped by Bill's brother, Walter, who lent his tooling skills and the black-painted machine was finished by Aunt Janet Davidson, who applied the pin-striping and original Harley-Davidson Motor Company logo on the tanks by hand.

Bill and Arthur had begun experimenting with 167cc (10.2 cu in) DeDion-type engines slotted into bicycle frames two years before, but it was not until Walter swapped his job in a railroad tool room in Kansas for a similar position at home in Milwaukee to work on the project that a prototype was finally completed. Working evenings and weekends in a friend's machine shop, they soon produced a second prototype, a larger single, 405cc (25 cu in), and received orders to produce two more. This second prototype was sold, and by 1908 it had passed through five owners and clocked up 60,000 miles (965.40km); in 1913 Harley-Davidson was advertising that it was still running and had clocked up 100,000 miles (160,900km).

Arthur and Walter's father, William C. Davidson, built a 15 × 10ft (4.6 × 3m) shed behind the family home, and production moved there in 1904, while Bill Harley went to Wisconsin University to study for a degree in engineering.

The backyard shed soon became too small, however, and it was doubled in size in 1905. Walter gave up his job at this time to become the first full-time worker. The pocket valve motorcycles that they built this year could achieve 45mph (73kph), although they still did not possess any brakes – the

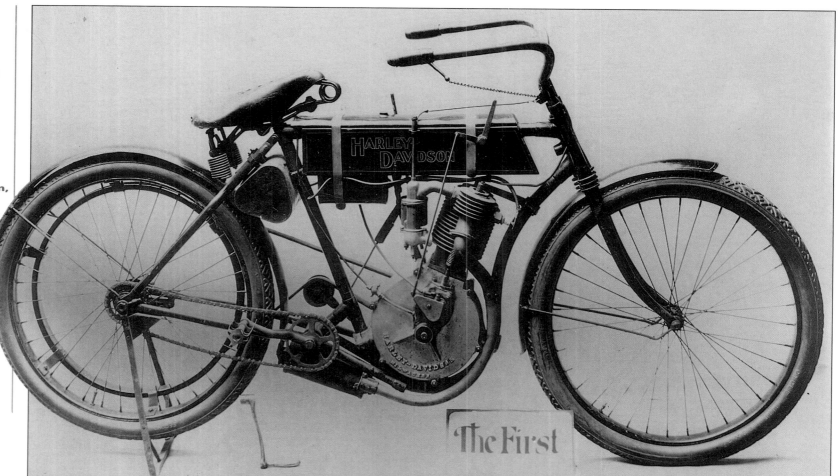

The first production model that Harley-Davidson made was this 405cc (25 cu in) machine. For the first three years black with gold striping was the only colour scheme on offer, until 1906, when the Renault Gray finish was available as an option. Notice the absence of suspension, and the oil tank attached on the top of the fuel tank.

The First

LEFT
The Silent Gray Fellow. The engine size of this model grew from 405cc (25 cu in) to 500cc (30 cu in) in 1909, and again in 1913 to 565cc (35 cu in). In 1911 the horizontal cooling fins on the cylinder head were replaced with vertical ones, and the curved downtube on the front of the frame was replaced with a straight one. From 1912 on the gas tank sloped downwards at the rear. 1914 was the last year that a belt drive was available on one of the singles. In 1916 rounder-shaped tanks were fitted and production of all singles ended in 1918.

rider had to pedal backwards to slow down.

The following year things expanded still further. The 'factory' moved into larger premises, the number of employees increased to six, production was up to 50, and the engine size was slightly enlarged. This 1906 model had an optional grey colour scheme, soon to become the only option, and it was known as the Silent Gray Fellow because of its baffled exhaust note and quiet running.

In 1907 William A. Davidson gave up his job to join his younger brothers, Arthur and Walter, and they formed a corporation with Bill Harley and sold shares in the company. These two families and their descendants retained a controlling portion of the stock, however, and it continued that way right up until the time the company was sold to American Machine Foundry (AMF) in 1969.

The company was always trying to improve its product, and during this year it fitted leading-link front forks, which for the first time gave a minimal degree of suspension to the 150 bikes that were produced. Comfort was further enhanced in 1912 with the introduction of the 'Ful-Floteing' seat, which sat on a long coiled spring inside the seat tube.

Walter Davidson's victory in the first Federation of American Motorcyclists (FAM) endurance run in 1908 (Chapter 3), made the press really sit up and take notice of what was happening in Milwaukee, and sales rose to 410. Meanwhile Bill

THE FIRST V-TWIN

ABOVE
A 1915 V-twin. From 1903 to 1920 Harley-Davidson made about 175,000 of their early singles and V-twins. They were also the only manufacturer to guarantee the horsepower on their motorcycles — on the twins it was an 11bhp minimum.

In 1909 Harley-Davidson introduced its first V-twin, but it was quickly withdrawn, suffering from problems with the intake valves. In addition, because there was no tensioning device on the belt drive, the leather drive belt was unable to handle the extra power. (It is surprising that a tensioner device was not fitted straightaway, for it was already available on the singles being made at the time.) The V-twin was re-introduced in 1911, with new mechanical inlet valves and a belt tensioning device that helped to transfer the power to the rear wheel. This also meant that the belt could be slackened so that the engine would idle and not die when the motorcycle came to a halt. Although this device worked well, a larger version of the V-twin (1000cc 61 cu in, previously 811cc/49.48 cu in) was available from 1912 with a chain drive.

This model, the 8-E, also featured Harley-Davidson's first clutch, which was installed in the rear hub. Thus the rider could pull away from a dead stop without having to pedal at all.

OPPOSITE
By 1914 there were almost 5000 Harley-Davidsons in use with the US Postal Service. Their ruggedness and reliability made them particularly suitable for deliveries in rural areas.

Harley had graduated from Wisconsin University and returned to start work on a new model.

Around this time, a 45 degree V-twin engine was being designed by most of the leading manufacturers, and Bill Harley set to work to produce Harley-Davidson's own version. By fitting a second cylinder, which added very little to the weight of the machine, the power could be simply more than doubled, and the extra cylinder would still slot easily into the bicycle-type frame that was used at the time.

Sales of Harley-Davidsons had increased to 12,904 by 1913, and the company felt confident enough to begin exporting its products, some years after most of its competitors had ventured abroad. Harley-Davidson signed an agreement with Englishman Duncan Watson, whose family was already well established in the import and export business, and he set up a shop in London and began planning the distribution of Harley-Davidsons throughout the rest of Europe. However, after only a small number of machines had been delivered, importation was interrupted by the outbreak of World War I in 1914 and did not resume until 1919. (Watson eventually became the Lord Mayor of London and was knighted – although this honour was not granted just because he was the first to introduce Harley-Davidsons to the UK.)

After a dramatic rise during the early years, sales of Harley-Davidsons in the US were beginning to level out, as the motor car became more accessible. With the introduction of the moving assembly line at Henry Ford's factory at Dearborn, Michigan, the Model-T cars were driving out of the factory gates in ever-increasing numbers. Although the cars were still nearly treble the cost of a good solo motorcycle, Ford started a scheme for instalment

plans to help more people to acquire their own four-wheeled transport.

When the USA became involved in World War I, the motorcycle manufacturers started producing machines for the troops. Harley-Davidson's main rival, the Indian Motocycle Company, switched its entire production to making military motorcycles, while Harley-Davidson contributed about half of its output. Indian's total commitment was made more for commercial reasons than an excess of patriotism, for this was at a time when its domestic sales were suffering. Harley-Davidson capitalized on Indian's absence from the domestic sales market by initiating a campaign to encourage its competitor's dealers over to its side. The campaign proved very successful and, after the war, Indian returned to civilian production to find its network of dealers depleted.

Harley-Davidson had been producing military machines in a matt olive green colour, but in 1917 this became the only colour available on the civilian models as well, although it was in a gloss finish. At first, it became highly fashionable to own a motorcycle in this colour, although people eventually became fed up both with it and a short-lived Brewster green variation. Yet it remained the only colour available for most of the next 16 years.

At the end of the war Harley-Davidson was able to take advantage of the increased efficiency that military production had imposed, and the company embarked on a programme of expansion. Construction started on a large extension to the factory, which, when completed, was to be the largest motorcycle facility in the world with room for 2,400 employees.

After peace was declared in Europe, Harley-Davidson re-established its export business to the UK, and soon afterwards Duncan Watson received

two of the 989cc (60 cu in) pocket valve racers that he had requested. These had been specifically ordered to try to win the title of the first motorcycle to exceed 100mph (160kph) in the UK. Five months later, on 28 April 1921, they did just that, after much attention had been paid to adjusting the handling characteristics of these bikes to take account of the bumpy Brooklands racing circuit. Douglas Davidson (no relation) attained 100.76mph (162kph). The following day, Bert le Vac, riding an 8-valve Indian, surpassed this with a speed of 106.52mph (171kph), but it could not take away the glory that had been captured for Harley-Davidson.

During the post-war lull at the beginning of the 1920s, sales plummeted to 10,202 in 1921, compared with the previous year's record high of 28,189. For the first time the company recorded a loss on the year. By this time the cost of a Model-T Ford motor car had dropped to the equivalent price of a sidecar outfit, and the three-wheel market was starting to lose sales. Just two years previously seven out of every ten Harley-Davidson's sold had a sidecar attached. To cut costs, the factory closed down for a month, wages were reduced and at the end of the year the Wrecking Crew racing team was disbanded. From now on the company only made machines available to individual riders.

Although sales at home had fallen, exports had remained consistent, although unspectacular. In an attempt to expand some of the smaller markets, Alfred Rich Child was despatched overseas. In conversations with Harry Sucher for his book *The Milwaukee Marvel*, Child relates how he arrived by sea at Cape Town, South Africa, in 1921, and then proceeded to ride north through the continent on a model J sidecar outfit. Roads and bridges were non-existent for much of his journey, and he was frequently carried through the jungle and across rivers by the natives. By the time he finally arrived in Cairo, Egypt, he had sold 400 motorcycles and set up several new dealerships.

Soon after he returned to Milwaukee in 1924, Child was sent across to Japan, where Indian was already established. Over a 13-year period Child imported 26,000 Harley-Davidsons and set up a

THE SPORT TWIN

In 1919, at a time when most US motorcycle companies were concentrating on producing large V-twins, Harley-Davidson tried to fill a gap in the middle-weight market by introducing the Sport Twin. The model was a 584cc (35.6 cu in), with cylinders horizontally opposed fore and aft, much like the Douglas, the British motorcycle produced in models of this type.

This bike came as a great surprise to the rest of the industry, not just because it was a departure from the V-twin, but also because it was rather dull to ride. With a top speed that might just make 45mph (73kph), the only advantages this machine had were the almost vibration-free ride that it gave and the accessibility it allowed to the engine for maintenance.

It was hoped that the Sport Twin would attract many new riders, but unfortunately for Harley-Davidson, Indian introduced its 600cc (37 cu in) Scout at the same time, and not only was it much more powerful it was also a V-twin, and even in those early days a V-twin was already regarded as *the* traditional American motorcycle. Although the Sport Twin did not appeal to the American public, it did enjoy some success in Europe before it was discontinued in 1923.

OPPOSITE
Harley-Davidson commercial outfits were used by many small businesses, particularly during the 1920s.

RIGHT
The story of the Japanese Rikuo is gradually being ommitted from the official history books.

licensing agreement that allowed several thousand machines to be built there under the name of Rikuo. Imports were suddenly stopped in 1937 when a military government came to power and raised import taxes to an excessive level. Although Harley-Davidson closed its office there, Rikuo continued to produce its version of Harley-Davidsons until 1959.

Back in the USA sales were beginning to recover and climbed again during the early 1920. After the withdrawal of the Sport Twin, only the 61 and 74 F heads were left in the line, until the company again ventured into the lightweight market.

After the low point in 1921, sales steadily increased throughout the decade and topped 22,000 in 1928. Harley-Davidson began a new era by replacing the F heads with new side valve engines – a 750cc (45 cu in) was produced in 1928 and a 1200cc (74 cu in) appeared late in 1929.

THE FIRST 74S – THE FD AND JD (F HEADS)

In 1921, to herald the new decade of optimism, the first 1200cc (74 cu in) V-twin was introduced. With its larger engine and greater power, it was more suitable for sidecar work than the 61, and it was also a rival to the Indian 74.

Like the 61 F and J models that had been launched two years previously and that had evolved from the first 1909 V-twin, these 1200cc (74 cu in) V-twins came in two versions – with magneto ignition (model 21FD) or generator (21JD). For the 1928 season a sportier version of the single cam J model was introduced, the JL (1000cc/61 cu in) and the JDL (1200cc/74 cu in).

This year also saw the arrival of the much-lamented 'two cammers', the JH (1000cc/61 cu in) and the JDH (1200cc/74 cu in). These 'two cammers' are still regarded by many as the best of the old V-twins that the company ever produced. However, although they were good for their time, they were not as perfect as some enthusiasts would have you believe, for they tended to overheat and break down when ridden at sustained high cruising speeds. Sadly, they were all discontinued in 1930 when the new model V was introduced, although the J model did continue in Class C competition until 1936.

The year 1928 is also notable as the year when Harley-Davidson fitted front brakes for the first time, although they were initially only on the big twins.

THE 350CC PEASHOOTER

In 1926 the 61 and 74 F heads were joined by another model, the 350cc (21.35 cu in), three-speed single, which became known as the Peashooter. It was available as a side valve (model A), an overhead valve (model AA) or with electric lighting (B and BA).

The Peashooter was conceived primarily for the export market, where there was a demand for lightweight machines, and it never achieved huge sales in the US. The OHV version, however, went on to win many competitions on short-track races and hill climbs in the newly formed 350cc (21 cu in) racing class, and it was also highly successful in races in the UK and Australia. Production was ended in 1930, but the machines nevertheless remained in competition for some years afterwards.

THE 750CC MODEL D

The 750cc (45 cu in) model D made a rather disappointing debut. It was supposed to compete against Indian's 101 Scout, yet its top speed of 55mph (88kph) was 20mph (32kph) down on its competitor, and its lack of appeal to potential buyers was compounded by an unreliable clutch and gearbox. It was taken off the market for a year while the problems were addressed and was relaunched in 1930 in three variations – the 15hp D, the 18.5hp DL and the 20hp DLD. The S models were geared for sidecar use (DS and RS).

The D model became the R (the R, RL and RLD) in 1932, when the generator was mounted horizontally, and there were many internal improvements – new flywheels, aluminium pistons, clutch, oil pump and so on – and it eventually became the W in 1937 when it was installed with a dry-sump lubrication system (the W, WL and WLD). The WLA version was reserved for the US Army and the WLC for the Canadian Army.

LEFT
A 1000cc (61 cu in), 61 J model. The 'olive drab' paint was the only colour available for most of the 1920s, although by 1928 Harley-Davidson had ceased painting the crankcases.

THE 1200CC MODEL V

The 1200cc (74 cu in) was virtually all new, sharing only a few parts with its F head predecessor. Despite weighing a massive 550lb (250kg), over 100lb (45kg) more than the F head, it was capable of a similar top speed, 80-85mph (129-137kph).

Although many 'two cam' enthusiasts still maintain that their mounts were faster than the new model V and that they were continually asking the factory to re-introduce them, few disagree that the old F heads were temperamental and unreliable. They may well have become better than the new V had they remained in production for more than two years so that there was an opportunity for their faults to have been ironed out. But the few surviving motorcycle magazines of the times were mindful of their small circulations and were not about to upset their main advertiser, so they toed the company line by promoting the new model V as a significant improvement over the 'two cammers'.

The V series eventually appeared in approximately 13 different guises. It is not possible to be more precise, for Harley-Davidson frequently introduced new models that had only slight cosmetic variations and that did not even appear in some catalogues. Nevertheless, the principal models were the standard V and the VL with slightly higher compression, and two versions equipped with a magneto, the VM and VLM. With the introduction of the dry-sump lubrication system in 1937, the V became the U (U, UL and so on), and just before that, a 1340cc (80 cu in) model was released (the UH and ULH).

As had been the case with the 750cc (45 cu in) D model, however, the V series was brought out prematurely after insufficient testing, and no sooner were the machines leaving the showroom than they were breaking down. The factory was inundated with complaints, and further deliveries of new machines were suspended. Within weeks a new engine design was in production with larger flywheels, new crankcases, valve springs, clutch plates and a modified frame among other improvements. Although dealers received a package to modify the motorcycles they had sold, they were not recompensed for the time it had taken them to rebuild their customers' machines.

Coming so soon after the problems with the initial 750cc (45 cu in) machines, this naturally caused great resentment among the dealers and some loss of faith among customers who had purchased one of the earlier troublesome motorcycles.

BELOW The sidevalve model V was produced between 1930 and 1937. From 1933 onwards a variety of Art Deco tank designs and new colours were introduced. This is a beautifully restored 1934 VL and sidecar.

THE 500CC MODEL C

This 500cc (30.50 cu in) side valve single was available with an optional hand clutch and was offered between 1929 and 1934. Like the 350cc (21 cu in) Peashooter, it was never a strong seller in the US, where a 750cc (45 cu in) V-twin was considered the minimum engine size by most of the public. It was first placed in the 350cc (21 cu in) Peashooter frame but later shared the frame of the 750cc (45 cu in) model and, unfortunately, also the clutch and gearbox that had caused so many problems. Tests were made on an over-head valve version of this single, but the plans for this were shelved.

Harley-Davidson had an on–off relation-ship with both the model C and the smaller 350cc (21 cu in) single. One year they would not be sold on the home market, only to re-appear two years later. The model CB appeared for only a short time, in 1933-4, and was really the 500cc (30.50 cu in) single in the 350cc (21 cu in) frame. At the same time, the 350cc (21 cu in) model B side-valve made a brief return.

Around 1930, 1200cc (74 cu in) machines accounted for approximately 53 per cent of sales, with 750cc (45 cu in) machines and singles representing 30 and 17 per cent of sales respectively. By the end of the 1920s over 40 per cent of production was exported, but soon after 1930 import taxes were raised in Australia and New Zealand, which were major markets for the singles, and the effects of the depression spread throughout the world. Exports quickly dropped to around 10 per cent of total production.

THE SERVI-CAR

The three-wheeled Servi-Car came out in 1932, and it remained in production right up to 1974. It was fitted with the 3-speed 750cc (45 cu in) D model engine, which powered a sprocket on the rear axle by a single chain. It was aimed at the light utility market, and it sold in large numbers to garages, which could send out mechanics on one with all their tools in the large box behind the rider, to work on a car and, if necessary, attach a tow bar and bring the car back. It was also employed by many police forces for traffic work. You can still see examples hard at work today in some Indian cities, where they have been converted into taxis and can now carry up to eight people, if not in comfort then certainly in great style.

The Wall Street crash in October 1929 devastated the US economy, and the motorcycle industry suffered along with the rest of business. Although sales dipped only slightly in the following year, they plummeted in the next two seasons, until, in 1933, they hit rock bottom at 3,703, then hovering around the 10,000 mark for the rest of the decade.

The demise of Excelsior in 1931 left Indian as the only domestic competitor to Harley-Davidson, and that company's sales also reached a low point in 1933, when just 1,657 machines left the factory gate. Not only had the lowering in the price of new motor cars been chipping away at sales, roadworthy second-hand Fords could be purchased during the Depression for under $100, while sales in Australia and New Zealand had virtually ceased because of prohibitive import tariffs that were being imposed.

At the depths of this economic gloom, in 1931, Harley-Davidson decided to brighten up its image and began chrome plating some of the small parts on its motorcycles. Larger items, such as wheel rims, could be ordered plated as well, but at extra cost. Finally, in 1932 customers were also allowed optional colours instead of the olive green or Brewster green that had been standard since 1917. Strangely enough, from 1926 new models could be ordered in a variety of bright colours, but this information was not mentioned in the sales catalogues nor in *The Enthusiast*, Harley-Davidson's own magazine.

The company had been sporadically producing both the 500cc (30.50 cu in) and 350cc (21 cu in) singles during the early 1930s, but in 1934 these were discontinued and production was limited to heavyweights. It was to be another 15 years until a motorcycle smaller than 750cc (45 cu in) was offered for sale.

THE KNUCKLEHEAD

After much research and testing – the company was not going to repeat the mistakes made with the side valves – the first overhead valve V-twin came on to the market in 1936.

This 61E model (1000cc/61 cu in), was, for reasons explained later in this chapter, the direct ancestor of today's successful Evolution engine.

The Knucklehead, as it came to be known (the shape of the rocker boxes on the right of the engine resemble the knuckles on the back of a hand), had many new features, some of which are still found on today's Harley-Davidsons, more than 50 years later.

For the first time Harley-Davidson installed a dry sump (oil circulating between the engine and oil tank) rather than a total loss oil system, with the horseshoe-shaped oil tank located underneath the seat and a four-speed constant mesh gearbox, and the whole engine was contained within a double-loop frame. The gas tank was welded instead of soldered and was in two halves, with the speedometer set into it (like the modern Softails). Further important developments to the 61E were a full valve enclosure in 1938 and a 1200cc (74 cu in) version – the 61F in 1941 – which could achieve 100mph (160kph).

Once Harley-Davidson was satisfied that early problems with the flow of oil had been sorted out and that the public had accepted the 61E – 2,000 were sold in its first year of production – an attempt was made to recapture the motorcycle speed record from Indian at Daytona Beach on 13 March 1937. Handling problems were rectified by stripping the fairing that enclosed the machine, and Joe Petrali went out on to the sand again, where he achieved a two-way average of 136.183mph (219.118kph) and triumphantly brought the title back to Milwaukee.

THE WLA 750CC

Harley-Davidson had already produced a military version of its 750cc (45 cu in) DLD for testing by the US Army before war broke out in Europe in September 1939, and shortly afterwards the company received an order to supply 5,000 military motorcycles for the British Army, because UK manufacturing capacity had been destroyed by German bombing.

Painted in gloss olive green, which was quickly replaced by matt olive green, this model was re-designated the WLA and was equipped with ammunition boxes, skid plate, rifle holster, luggage rack and blackout covers on the lights. Mechanically, these bikes proved to be virtually indestructible. With their low compression, low speed – 50mph (80kph) was the maximum – and substantial torque, they proved, during service with the Allies' armies, to be highly reliable and capable of enduring sustained abuse.

BELOW
The military version of the WL model (the letter 'A' after the WL designated the army version). The Canadian Army version, the WLC, can be easily distinguished from the US Army version, the WLA, as it was equipped with the stand on the front wheel, and invariably came with a passenger seat.

THE PANHEAD

The Knucklehead was replaced by the Panhead in 1948, but it retained the model designations E and F for its 1000cc (61 cu in) and 1200cc (74 cu in) versions.

The introduction of the Panhead started a trend that continues to this day – the modern Evolution engine was developed by fitting a new top end on to the existing Shovelhead engine; the Shovelhead had a new top end fitted to a Panhead engine; while the Panhead engine featured a new top end fitted to the bottom half from the Knucklehead. The name Evolution, therefore, derives from the engine's having evolved from the Knucklehead. Obviously, some small but important changes were made over the years to enable these transitions to be made, which, for reasons of space, cannot be detailed here. Full details of these changes, however, are included in the excellent books by Allan Girdler or Jerry Hatfield.

The most striking visual difference between the Knucklehead and the Panhead was the aluminium heads, which were said to resemble upside-down baking pans. These enabled the engine to run cooler and helped to keep the engine oil tight. Despite many improvements, some of the problems with the all-iron Knuckle-head were never resolved, and they could still at times suffer from overheating and oil leakage around the rocker boxes. Inside the Panhead, many refinements had been made. Hydraulic lifters removed tappet noise, a larger pump improved the flow of oil and there were no external oil lines.

Further important developments on the Panhead were made in 1949, when telescopic front forks were used on the Hydra Glide, in 1952, when the optional foot gear shift was added to the FLF, and in 1953, when the distinctive trumpet-shaped air horn was first used. In 1956 high lift cams were introduced; in 1958 rear suspension was used on the Duo-Glide, and in 1965 an electric starter was developed for the Electra Glide.

The standard 1000cc (61 cu in) E and 1200cc (74 cu in) F had been dropped by 1952, and the rest of the 1000cc (61 cu in) E models, EL, ELF and ES, had disappeared by 1953.

ABOVE
A 1957 FL that was purchased from the Vietnamese government in 1989, exported and restored.

BOTTOM LEFT
A 1964 Duo-Glide.

BELOW
A 1959 Duo-Glide.

Unions had become accepted by all the motor car manufacturers, and, in 1936 unionization reached the gates of the Juneau Avenue factory. After much negotiation, the workforce's decision to introduce a union was accepted, and it was formally admitted in 1937. Shortly afterwards there was an increase in wages, and the price of motorcycles was raised accordingly.

One of the four founders, William A. Davidson, died in 1937, his death marking the beginning of the end of an era. Walter Davidson died in 1942, and William S. Harley died in 1943. The surviving member of the original quartet, Arthur Davidson, remained an active director until he was killed in a car accident in 1950.

Many of the founders' sons and relatives were employed at the factory – by 1939 20 members of the two families were stock holders – and the title of President was passed to William Herbert Davidson, the son of William A. and father of Willie G., and he remained the head of the organization until 1973.

When it became clear that, sooner or later, the US would become involved in the war, production at Harley-Davidson was put on a full-time basis, with the factory working both day and night, concentrating on WLAs. Later in 1941 the company issued a bulletin that, instead of announcing the programme for the following year as it would normally have done, told dealers that because of shortages of materials and the concentration on building military vehicles dealers would be able to order only one motorcycle from the factory. Three months later, after the Japanese attack on Pearl Harbor in December 1941, America joined the war and most of the dealers were not even supplied with one model, as production was limited to military WLAs.

THE XA 750CC

During this period the US Army was testing the XA, of which it had asked Harley-Davidson to build 1,000. The XA was a copy of a BMW 750cc (45 cu in) side valve twin, with horizontally opposed cylinders and shaft drive. It was designed for use in the desert, because a shaft-drive motorcycle would be more suitable for sandy terrain than a chain-driven machine. However, the four-wheel drive jeep entered production at this time and was deemed more suitable for this purpose, so the XA project was terminated.

By the end of the war Harley-Davidson had built almost 90,000 WLAs for the Allied forces and 20,000 alone, the WLC, for the Canadian Army. However, many of these had not even reached the Army and were still in their crates, and to off-load this surplus the company sold them to the public still in their military trim. While the factory had been totally committed to supplying the military – 29,000 machines had been produced for the Army in both 1942 and 1943 – the civilian rider had had to contend with a speed limit of 35mph (56kph), synthetic rubber tyres, gas rationing and a severe shortage of spare parts.

When peace was declared in 1945 the US stood alone as the only nation capable of major industrial production. While the rest of the world was rebuilding after devastation caused and denying US companies an export market with high import tariffs, Harley-Davidson was having to contend with both a nationwide shortage of materials and motorcycles imported from the UK. Out of a total of 30,000 new machines sold in the US in 1946, one-third came from the UK, and many of them were singles.

By 1947 consistent supplies of raw materials had been re-established, and Harley-Davidson could resume full production. These were boom years for the company as sales climbed to an all-time high of 31,163 in 1948. Although the 1200cc (74 cu in) side valve was officially discontinued at this time, it was still available on order for a few more years.

By 1950 US manufacturers still had to contend with high taxes on their exports, while imports into the US had only low taxes imposed on them. In consequence the US was flooded with goods made with cheap labour from all over the world.

The wide availability of overhead valve vertical twins, such as those made by Triumph, BSA and Norton, meant that the US public could buy UK-made bikes that were almost half the capacity and weight of their big V-twins yet were cheaper and could match them for performance. Both dealers and the public were switching over to the imported machines in ever-increasing numbers.

Harley-Davidson, whose sales had been declining, fought back by asking the US government to increase substantially the import tax to restrict the flood of foreign motorcycles coming into the country. The importers of the UK machines organized themselves to oppose any increase in the import tax at the Senate Committee hearings, which were held in 1951 and among them was Alfred Rich Child who was now importing BSAs into the US. The hearings granted judgement in favour of the importers, and Harley-Davidson had to compete for sales at a disadvantage.

Indian was also suffering financially at this time. The heavyweight Indian Chief was being

LIGHTWEIGHTS

Harley-Davidson had received many requests from their dealers for a 350cc or 500cc (21 or 30.50 cu in) bike to compete with the UK motorcycles that were being imported in increasing numbers. The company responded by saying that it did not have the capital to develop a new model. Instead, it introduced a three-speed 125cc (7.6 cu in), which was a copy of the German DKW. Soon afterwards, the UK manufacturers BSA produced its own copy, which was called the Bantam.

This little two-stroke was aimed at school boys and for light utility work and, with the aid of a finance package, could be purchased for $5.50 a week.

It was the first of several 124-175cc (7.6-10.7 cu in) lightweights, which were produced from 1947 onwards. Many of these had rather cute names – Topper, Bobcat, Hummer, Ranger, Pacer and Rapido – and the 50-65cc (3-4 cu in) machines were known as Sport and Shortster. Some of these lightweight models were made at the Aermacchi plant in Italy after Harley-Davidson had taken it over in 1960.

The 125cc (7.6 cu in) Hummer that appeared in the late 1950s. This was a slightly updated version of the first lightweight that Harley-Davidson had made in 1947.

produced in only limited numbers, while the company concentrated its efforts on designing overhead valve lightweights to compete with the UK imports. Soon after construction started on these small bikes, however, production costs rose dramatically, and Indian was unable to raise the finance to continue. Many of its dealers had closed down, while others had taken on UK-made machines or switched to Harley-Davidson. The Indian Motocycle Company was divided up and production was taken over by Atlas Titeflex Corporation, which produced about 3,000 Chiefs in the short time that it owned the company, until it finally closed down in August 1953.

Harley-Davidson was now the last US motorcycle manufacturer left.

Not only did Harley-Davidson have to compete for the domestic motorcycle market with UK-manufactured cycles, but the Japanese had now arrived with some small bikes, and were slowly starting to undermine the US market.

THE K MODEL

At the time of Indian's demise, Harley-Davidson seems to have taken a step forwards, and a step sideways at the same time.

Production of the WL was finally terminated in 1951 after 22 years of service. Its replacement, the model K, appeared in 1952, looking very much like its modern European competitors, with a swing arm, rear suspension, telescopic front forks, left-hand clutch and right-foot gear change. It was, however, powered by a side valve engine of unit construction that was based on the 750cc (45 cu in) V-twin, which had first appeared in the 1920s. With a top speed of 80mph (129kph), it was far slower than its smaller overhead valve competitors. A sportier version was quickly developed by blending the K with the racing KR to produce the KK.

To improve performance the motor was enlarged from 750cc (45 cu in) to 883cc (55 cu in) two years later, and this KH model could achieve a more respectable 95mph (153kph). The KHK derivative was the 'hotter' street bike.

The K was regarded as a disappointment when it was first released. It arrived just when buyers wanted a US motorcycle that could compete with the smaller 500cc (30.50 cu in) Triumphs and BSAs, and although later versions sold better, the bike was generally regarded as a stopgap until the factory could produce something that could really blow everything else away.

It says a lot for the talents of Harley-Davidson tuners Tom Sifton and, later, Dick O'Brien that the racing version of the K, the KRTT, won at Daytona on its first outing there in

1953, finishing far ahead of the second bike home. It won its last race there in 1969, too, and to rub it in, Cal Rayborn even lapped the rest of the field.

Although the racing KR had to stay competitive for 18 years until it was replaced by the XR-750 in 1970, the road-going model K lasted for only four years – until in 1956, when an important development occurred.

BELOW
A 1954 883cc (55 cu in) KH model. While the KH was significantly better than its 750cc (45 cu in) model K predecessor, the sportier KHK was the only version that offered serious competition to the English bikes on the street. Not that many KHKs were sold, and original bikes in good condition are extremely rare.

THE SPORTSTER

In a way that was reminiscent of the development of its big twins, Harley-Davidson designed a new 883cc (55 cu in) overhead valve top end and fitted it to the K engine in 1957. This new XL engine was slipped back into the K frame, christened the Sportster and, with a few other changes, sent out to repel the invasion of UK motorcycles.

It was a great success right from the start, achieving twice as many sales as the K in the previous year. True to form, Harley-Davidson introduced a higher compression version – 9:1 compared with the XL's 7.5:1 – and fitted larger valves the following season, when they called it the XLH.

Even this boost in power was not enough for many, and the factory was inundated with demands from dealers for an even faster version. For once the public got what it wanted, and the XLCH arrived late in 1958. This sporting version was sold with high-level pipes, a magneto ignition, semi-knobbly tyres and the small peanut tank that is, today, standard on Sportsters. The standard XL power output was 40bhp, but the XLCH produced 55bhp and was capable of exceeding 120mph (193kph) and ¼-mile (400m) times of 14 seconds. This model, unsurprisingly, was soon outselling the standard XL, and it was many years before anything else appeared that could match its performance.

The XLCH was tamed over the years, and despite an increase in size to 1000cc (61 cu in) in 1972, the top speed was slowly reduced, and buyers switched their attention to the now almost

as fast but more civilized XLH. Production of the XLCH finally ended in 1979.

Other variations were the XLX, a very basic 1000cc (61 cu in) Sportster, the XLT touring version and the XLS Roadster, which was styled after the FX Lowrider. There were also two 'hot' Sportsters. The XR-1000, which was styled like the successful racing XR-750 with twin carburettors on the right and the exhausts running high along the left side, was basically a standard XL 1000cc (61 cu in) engine with the XR-750 heads engineered by Jerry Branch. At $6,995 it did not find many buyers, and those who were interested in performance invariably purchased the

ABOVE
The XLCH was initially intended for off-road use and came with these high-level exhaust pipes. Later versions had the short dual pipes that became traditional on Sportsters. The bike was blessed with a strong gearbox and a clutch that seemed indestructable, even when drag racing.

$3,995 XLX and fitted the hotter twin carburettors, cams, alloy heads and exhausts themselves – and still had plenty of change left over.

Like the XR-1000, the XLCR that was introduced in 1977 lasted for only two years. This all-black Cafe Racer could reach 120mph (193kph), but did not sell to the sporting rider and was not considered a 'real Harley' by those who think they know best.

The XL Sportster is still with us, however, more than 35 years after it was first introduced.

BELOW
Only 3000 XLCRs were ever made. The new frame had a triangular rear subsection and suspension that was derived from the racing XR-750. Both the frame and the two-into-one exhausts had been passed on to the rest of the XL range by 1979.

ABOVE
The XR-1000 gave out 70bhp in its standard form off the showroom floor. With higher compression pistons and a high lift cam 80bhp could be reached, with potential to go still further and exceed 100bhp — at extra cost.

AERMACCHI

In 1960 Harley-Davidson purchased a half share in the struggling Italian Aeronautica Macchi Company, which produced motorcycles. This enabled it to import and sell new, lightweight motorcycles that could be built cheaply and which had already been proved successful in Italy.

Soon after the deal was made, the first of the Aermacchis arrived in the US. It was renamed the Harley-Davidson Sprint. This four-stroke 250cc (15.24 cu in) single could top 70mph (113kph), and by 1967 it was Harley-Davidson's biggest seller, while the 350cc (21 cu in) version that arrived in 1969 and could exceed 90mph (145kph), sold almost as well. In 1972 these four-strokes were replaced by two-strokes from the Aermacchi factory in an attempt to keep up with the competition, but by 1977 the Japanese had taken over the market for two-strokes and Aermacchi was sold back to the Italians. Its new owners subsequently renamed the company Cagiva and turned it into the largest motorcycle manufacturer in Europe.

The racing versions of the model C Sprint – the CR, CRS and CRTT – were frequent winners in the short-track races in the US, while the two-stroke versions, the RR 250 and 350, were ridden to four world championships on the Grand Prix circuit by Walter Villa.

LEFT
The four-stroke Aermacchi Sprints came in both 250cc (15.24 cu in) and 350cc (21 cu in) engine sizes. In 1968 a tuned 250cc (15.24 cu in) version of this single exceeded 177mph (285kph) at Bonneville.

RIGHT
The 165cc (10 cu in) Topper scooter was produced from 1960–65 during a brief period when scooters were fashionable, particularly with students.

BICYCLES, GOLF CARTS AND SNOWMOBILES

Harley-Davidson had frequently diversified and built non-motorcycle products. Beginning with bicycles in 1917 (which sold poorly and were discontinued five years later), the company went on to build generators, marine winches and outboard engines.

The three-wheeled golf carts were an extremely successful diversification, which at one time accounted for almost 20 per cent of annual sales. They were manufactured in electric- and gasolin-powered versions.

A Snowmobile was produced from 1971 to 1974. It was powered by two-stroke engines and was available in 24.29 and 26.42in (62 and 67cm) versions, which enjoyed a brief popularity until the novelty wore off.

THE SHOVELHEAD

shift was considered to be standard and was available until 1973); the B was fitted with an electric starter; the P was the police model; and the T had a five-speed gearbox and a new rubber-mounted frame.

Differentiating between a FLHFB and a FLP may seem complicated, but it is quite simple really. However, the B designation was dropped in 1970, when the electric starter became standard equipment, and the F disappeared in 1973, when the hand shift was no longer offered. So a 1974 FLH would be basically the same model as a 1969 FLHFB.

LEFT
1976 FLH.

BELOW
This late 70s Shovel-

head features a tank logo based on the Art Deco design that first appeared on the 1936 Knucklehead.

The Panhead gave way to the Shovelhead in 1966. This was topped with alloy rocker boxes, which were very similar to those that had first appeared on the iron head Sportster in 1957. Its nickname came about as the covers were said to look very similar to the back of a shovel.

The later Panheads had external oil lines, and these were continued on the Shovelhead. For the 1970 model the generator was exchanged for an alternator inside the primary case, with a new cone-shaped timing cover, giving rise to the names Generator Shovels and Alternator Shovels (Cone Motors). In 1978 the 1340cc (80 cu in)

version was introduced.

The Shovelhead continued to have the same designations on the basic models that had been applied to the Panheads – that is, the FL and FLH – but one of the most striking differences between the new Shovelhead FLH and its 1952 FL Panhead predecessor was the 200lb (over 90kg) increase in weight. The FLH weighed in at a massive 783lb (355kg) when fuelled up. There were many variations on the now standard 1200cc (74 cu in) FL. The H was boosted another 6hp to 60bhp and was loaded with the touring extras; the F featured the foot shift (the hand

By the mid-1960s Harley-Davidson's share of the domestic motorcycle market had shrunk, and only 3 per cent of its production was exported. Survival could be assured only by a large injection of capital, and so, in 1965 and 1966, it went public and sold shares in the company, although more than 50 members of the two founding families continued to retain a majority share holding. Even this was not enough to ensure the continuation of the company, for two years later it offered itself up for a merger or take-over.

Two companies expressed an interest – Bangor Punta and the American Machine & Foundry Company (AMF). Both companies had huge assets and holdings in a wide variety of industrial and leisure products, and they fought in and out of court to acquire control. The Harley-Davidson directors favoured AMF, and eventually the stockholders accepted the offer and AMF took over on 7 January 1969, thus bringing to an end the family-run business. Nevertheless, the board of directors, most of whom were still Harleys and Davidsons, remained in office.

It was not long before the inevitable changes were introduced, and the management style inherent in large corporations led to AMF becoming deeply resented. Production was speeded up to a point where quality control suffered. A new plant was opened in York, Pennsylvania, and the principal components were shipped there from the Milwaukee factory for assembly. The workforce became demoralized by the increased pace at which it had to work and by the sub-standard product that was leaving the factory. When the assembly-line was moved to the York, Pennsylvania, plant, the union called a strike over the loss of 700 jobs because it had been promised that there were to be no lay-offs.

LEFT
The genial Willie G, design director at Harley-Davidson. With his team he has been responsible for most of the styling and advances in the company's 'bikes' designs over the last 20 years. Many memorable creations include the Super Glide, The XLCR Cafe Racer, The Wide Glide, The Sturgis and the Softail.

William G. Davidson, Willie G., the grandson of founder, William A. Davidson, was working in the automotive design department at Ford when he received a call from his father, William H., the company president, to join the Harley-Davidson design department in 1963. By the end of the 1960s he was working on a design for a new model that would fill the gap between the Sportster and the Electra Glide. What emerged was a motorcycle that blended the frame, rear suspension and 1200cc (74 cu in) Alternator Shovelhead engine from the FLH with the trimmer front fork and wheel from the XL Sportster. Stripped of the FLH's boxes, fairing, electric starter and huge battery, it was almost 200lb (90kg) lighter.

This was the 1971 FX Super Glide: 'a true watershed in US riding.'

Harley-Davidson was an enthusiastic user of fibreglass at this time. The Tomahawk plant that built the bodies for the golf cart and fairing for the Electra Glide were utilized again on this project. The designers created a large, all-in-one stepped seat base and rear fender, with the rear light set into the tail, and it was painted red, white and blue and this paint scheme was called 'Sparkling America'. Unfortunately, fewer than 4,000 models were sold, and the fibreglass was replaced the following year for a more conventional metal seat base and separate fender. Just like a later Willie G. design – the short-lived XLCR Cafe Racer – people only recognized how good it was after it had gone, and both machines have since become highly sought after.

The FX became a huge seller and was a slight nod in the direction of the custom and chopper market that Harley-Davidson had ignored for so long. Additional letters after FX that appeared over the years were as such: E (electric starter), R (five-speed gearbox) and DG (disc glide).

By the end of the 1970s Harley-Davidson had come alive to the fact that it was losing revenue to the customizers, and it started to give so many nods to the custom market in rapid succession that it almost became a nervous twitch.

THE FX

THE FXS LOW RIDER

The wheelbase was lengthened, while the seat was lowered by over 2in (5cm) and the footrests were moved further forward. It was also equipped with siamesed 2-into-1 exhaust pipes, special paintwork, mag wheels and flat drag handlebars.

THE FXEF FAT BOB

This was fitted with the fatter 5 gallon (23 litre) fuel tanks, which had appeared on the first FX but had been replaced on later models by a slimmer 3.5 gallon (16 litre) tank. It also had a shorter, 'bobbed' rear fender.

THE FXWG WIDE GLIDE

Basically a Fat Bob with slightly longer and wider front forks and a 21in (53cm) front wheel, 2in (5cm) larger than the standard FX, what really set this bike apart were the flames painted on the tanks to imitate the custom trend at the time.

THE FXB STURGIS

This was a Low Rider, but with a black chromed engine and black paint work, and it was powered by belt drive instead of chains.

Some of these models had a 1200cc (74 cu in) engine, but they were all eventually offered with a larger, 1340cc (80 cu in) engine, which had just been fitted to the top-of-the-range Electra Glide.

RIGHT
**Two Sturgis models.
On the left is the
limited edition 1990
model that was
reintroduced to
celebrate the 50th
Sturgis rally, and on
the right is the
original 1980 version
that has had some of
its black parts
replaced with
chrome accessories.**

OPPOSITE
**The 1971 FX Super
Glide with the
fibreglass seat. A
similar seat unit was
also offered as an
option for the
Sportster that year,
but this attracted few
buyers. This giant
leap forward in
design for Harley-
Davidson was ready
to go into production
in 1967. However,
only when AMF had
taken over was the
go-ahead given.**

Many of the dealers who remained loyal to Harley-Davidson during these difficult years, recall that it was standard practice to take apart and rebuild every new machine that arrived to discover and rectify all of the delivery faults. The public began to refer to two different types of Harley-Davidsons: the AMF and the pre-AMF bikes – and it wanted only the latter.

Before the take-over by AMF Harley-Davidson had had the usual strengths and weaknesses of many long-established family-owned companies, including a strong sense of its own history, a remarkable ability to make do and a frequent lack of foresight for the future. AMF, on the other hand, had only a strong sense of what it thought was right for the future and what the bottom line should be. Surprisingly for such a large, successful company, it got it wrong too many times.

Despite the problems, the early 1970s were boom years for motorcycle sales, and AMF doubled production from 27,000 to 60,000 within three years. However, as we have seen, quality control could not keep up with the numbers of bikes that were coming off the assembly lines. Vaughn Beals, who joined the company in 1975 as a vice-president and who headed the buy-back in 1981, recalls (in *Well Made in America* by Peter C. Reid), a spot check that he instigated on the new XLCRs that were about to leave the factory after they had been passed for shipping: 'the first one hundred Cafe Racers that came off the line ready to ship, cost us over a hundred thousand dollars to fix before they could pass inspection.

Meanwhile, the marketplace was changing. Honda had spearheaded the invasion of Japanese motorcycles into the US when it introduced a small moped in 1959. Other Japanese manufacturers soon followed. Initially they concentrated on the

THE XL SPORTSTER

Still around after 35 years and available in four models with 883 and 1200cc (55 and 74 cu in) engines, the XL Sportster continues to be a popular introduction to Harley-Davidsons for many people. It now comes with the five-speed gearbox and belt drive on the 1200cc (74 cu in) and 883cc (55 cu in) de luxe that are fitted on all of the 1340cc (80 cu in) models.

LEFT
A 1989 XL-1200. The original exhausts have been replaced with accessories that do away with the balance pipe that obscures the beautiful V-shape of the engine.

light- and middle-weight market that Harley-Davidson had chosen to ignore for so long. But by the time AMF held the reins, Japanese companies were flooding the market with large superbikes that were cheaper and more reliable and that could out-perform anything Harley-Davidson offered.

By now there was an almost universal dislike of AMF among many Harley-Davidson employees, and an almost universal loathing of the AMF product outside the company: the machinery was inefficient, the motorcycles it produced were out-dated, over-priced and often unreliable, and the company operated on a very slim profit margin. By 1980 AMF wanted out.

Fortunately, 13 Harley-Davidson executives wanted in, and, led by Vaughn Beals, they managed

THE V2 EVOLUTION (EVO)

This was the first option in the new range, and it came out in 1983. This Evo engine – it was nick-named the Blockhead at first in the time-honoured tradition, but the name failed to catch on – was so called because it had evolved out of the Pan- and Shovel- and Knuckleheads and had a new top end fitted to the bottom of the old Shovelhead engine.

It first appeared on the FX and spread across the rest of the range over the next three years. Today Harley-Davidson makes only two different types of engine, both of them Evos. The Sportster, which internally comes in two sizes but which both look the same externally, and the 1340cc (80 cu in), which is fitted to the rest of the range. Together they appear in the 18 motor-cycles that Harley-Davidson now sells.

THE FXR LOW RIDER

The FXR is based on the old Low Rider with a conventional swing arm and rear suspension, but now the motor is isolation mounted to smooth out any vibration. It is available in six models, mostly distinguished by cosmetic variations, ranging from the touring FXRT to the FXLR Custom.

to raise the finance to buy back the company and survive. To raise more capital, they went public in 1986 and sold shares in the company. Anyone who wants the full details on the buy-back story should read *Well Made in America* by Peter C. Reid, which relates the long and torturous financial struggle that brought Harley-Davidson back from the brink.

It is, perhaps, worth making the point that, if AMF had not taken over Harley-Davidson, Harley-Davidson would have been taken over by Bangor Punta, which would, in all likelihood, have stripped it of its assets, and most certainly Harley-Davidson would not be producing motorcycles today. For all its faults, AMF arrived with the in-tention of turning Harley-Davidson into a profit-able company, and it poured a great deal of money into it to achieve this end. Although it did not suc-ceed, it is an indisputable fact that Harley-Davidson has managed to survive and become a successful company today only because of the modernization, expansion in production and development of the Evolution engine that AMF initiated.

In the late 1970s Harley-Davidson had be-gun to develop two new ranges of engines. The first was based on the line of engines that went back as far as the 1936 Knucklehead. The second

ABOVE
The modern Softail has the speedometer and instruments set into the twin Fat Bob tanks. This arrangement was first used on the 1936 Knucklehead, and it leaves a clean uncluttered look around the handlebars.

was a development of a completely new range of two- to six-cylinder water-cooled engines from 500 to 1500cc (30.50 to 92 cu in). This was the NOVA project, which was eventually shelved in 1983.

Between 1980 and 1982 Harley-Davidson had to lay off half of their workforce, and the situ-ation had never looked grimmer. In 1982, and not for the first time, the company appealed to the US government to apply higher import tariffs to the Japanese heavyweight motorcycles that were inun-dating the country and slowly overwhelming the company. Fortunately, this time the government was listening, and in 1983 tariffs of up to 50 per cent were imposed on imported Japanese motor-cycles of over 700cc (43 cu in). Honda and Kawasaki switched to making more heavyweights in their US plants instead, but the import duties gave Harley-Davidson some breathing space in which to begin the long haul back.

By 1987 the company was in good shape, and it even took the unprecedented step of asking that the tariffs imposed on their competitors be lifted as it was now strong enough to compete on level terms. Although the tariffs had, in fact, become largely ineffectual, the move created an enormous amount of favourable publicity for the company and was a positive affirmation to the world that Harley-Davidson was back on top.

THE FL TOUR AND ELECTRA GLIDES

The five Glides also have the engine isolation mounted, and they differ from each other only in the quantity of fibreglass and gadgets that are added on – the top of the line models even come with a cigarette lighter.

THE FXDB DYNA GLIDE STURGIS

The limited edition Sturgis first appeared in 1980. It was basically an all-black Super Glide with a belt drive installed instead of a chain. So successful was this that belts gradually replaced chains across the entire range. The Sturgis was brought back in 1990 and installed in a new chassis with isolation engine mounting. For 10 years Harley-Davidson bikes had a balance pipe running between the two exhaust header pipes, and this cut across, and to this writer's eyes at least, spoilt, the distinctive V-shape of the engine. With this model, though, the company has found a way to tuck it out of sight.

THE FXS SOFTAIL

BELOW AND INSET 1990 FXSTS Softail Springer. The rigid-looking frame and modern Evolution engine create a perfect blend of the old and the new.

The four Softails take their styling from an earlier age with their bobbed or traditional rear fenders, chrome wrap-around oil tank and a rigid-looking frame (the suspension is cunningly hidden). The FLSTC Heritage Classic comes with a transparent windshield, studded saddle bags and front fender, which imitates the look of 'dressers' from the 1950s and just oozes nostalgia, while the FXSTS Springer takes this look back even further with the sprung front forks that last appeared on a Harley-Davidson in 1948.

Customizing

Customizing is older than Harley-Davidson itself, although the alterations that many of the early riders made to their bikes were not always done to improve the looks of the machine. In the early days it was possible to buy all the components you would need to construct a motorcycle from mail order companies. When the postman delivered them, all you had to do was bolt them all together in your shed, paint a name on the gas tank, and, lo and behold, you had your own motorcycle company.

This is how many of the early manufacturers began and also why so many went out of business so quickly. Many of these backyard creations were unreliable, and even if they lasted for any length of time, they would be worn out after only a few thousand miles. Anyone who purchased one of these machines soon came to regret it, and needed to be adept at making parts to keep it running.

The Silent Gray Fellow produced by Harley-Davidson was a notable exception to all of these, and it gained an enviable reputation for its reliability and quiet exhaust note. Many of the riders who purchased the Harley-Davidson enjoyed the reliability, but even then they would poke a metal rod up the exhaust pipe to knock out the baffles to give the exhaust note a more noticeable sound. This tradition continues today, as many riders return home from the showroom with their new pride and joy and do exactly the same thing. The exhaust note need not be offensively loud – just enough to know you are riding a V-twin, and, as the stickers say, 'loud pipes save lives'. In those early days though, motorcycling was struggling to remain respectable, and Harley-Davidson advertisements would ask owners not to indulge in this practice 'lest the entire wholesome sport of motorcycling be banished'. Fortunately it has not been. Yet.

LEFT
Close-up of a show bike built for winning prizes, rather than for any practical purpose. The cooling fins have been cut off at the corners to give a more striking shape to the engine and all available surface area has been either engraved or gold-plated.

OPPOSITE
A French customized Softail, with a typically idiosyncratic French colour scheme that spills over on to the engine.

Before long Harley-Davidson were offering a choice of different-coloured pin striping on its grey machines. There is much debate as to the actual choice of colours available then because of a lack of surviving documentation, but today you can order a new Harley-Davidson in your own combination of colours.

By the late 1920s it had become fashionable for club members' bikes to be fully dressed with windshield, saddlebags, crash bars and spot lights at the very least, and many owners went beyond this by attaching as many extra lights and chrome accessories as possible. Club weekends invariably included a competition, often sponsored by the local Harley-Davidson dealer, for the best 'dresser' with the most extras fitted. The practice continues today, and it would seem that for many riders the excessive loading of as many lights and fixtures as possible is the only way to go.

After World War II many riders on the west coast of the US, who were not part of the 'dresser' scene, started to create their own look. Many of these riders chose the big V-twin as their mount, and Harley-Davidsons predominated. These riders would buy one of the dressers but would discard all that was superfluous to a basic motorcycle, sacrificing comfort for a basic, stripped-down look. Windshields, saddlebags, large seat, crash bars and the front fender would all end up in a pile in the back of the garage, while the rear fender would be 'bobbed' by cutting off the hinged section, a practice that gave rise to the name 'California Bobber'. After shedding all this weight, not only did the bike look raw and lean but its performance was greatly improved.

The next step came by way of the drag strip, in a move to imitate the elongated machines that tore along the ¼ mile (400m). During the early

A Knucklehead 'Bobber'. Apparently this bike was put together in the 1950s and remains virtually unchanged.

The latest creation by Arlen Ness. The engine capacity is over 2000cc (122 cu in) and it is fed by four dual Dell 'Orto carburettors, nitrous and supercharged.

A Buell RS-1200. Erik Buell makes frames for racing bikes and also produces a number of these hand built street machines every year. Restricted by emission laws for manufacturers, the Sportster engine has to be standard when sold.

1960s some riders experimented with telescopic forks, bolting on front ends that were longer than the standard. Some say that this practice started even earlier, when springer forks were cut in half and welded back together with extra tubing added. Whichever came first, both practices resulted in a raising of the front of the motorcycle, which also made it rather unstable.

A more practical imitation of the drag racers soon appeared. The frame was raked and stretched to accommodate the longer forks, while the engine and rider were kept close to the ground for stability. A fat rear tyre and a small, skinny front tyre would be fitted, and tall 'ape hanger' handlebars and sissy bar were attached. The Fat Bob gas tanks would join the pile of unwanted parts in the back of the garage, to be replaced by a small 'peanut' tank. Once all the coats of paint were dry, off you could cruise on something that was truly different.

These early choppers were made at home, with parts bolted together from various sources, but it was not long before shops appeared that specialized in making parts and putting together bikes like this. These shops met a real demand, for few authorized Harley-Davidson dealers would service or repair these creations.

One of these innovators, if not *the* innovator was Arlen Ness. At first he made modifications to his own bike, purely to please himself, but he was inundated by friends who wanted him to make similar alterations to their own motorcycles. Ness' creations have been copied by many, and today he is still one of the leaders in this field, manufacturing custom parts (many of his accessories are even available from Harley-Davidson dealers in the US) and putting together some of the most radical and adventurous bikes around.

A whole industry has developed to cater for

those who want to make their motorcycles unique. Today you can build a Harley-Davidson solely from parts supplied by aftermarket suppliers and made in the Far East – pistons, cases, gearboxes, frames, wheels . . . the lot. A host of quality aftermarket suppliers offers a huge range of custom and performance parts, among them are Delkron, Jammer, Custom Chrome, Cycle Shack, Performance Machine and Paucho.

In the 1960s Harley-Davidson regarded the custom scene with great disdain. However, it reached the point where the company could no longer afford to ignore it, and when Harley-Davidson finally realized how much business it was losing to the custom market, it responded with its own 'factory custom' motorcycles.

The Super Glide, introduced in 1971, was the first tentative step in that direction, but it was the Low Rider, which appeared in 1977, that made a more positive statement. With its low seating, forward foot controls and forks that were slightly raked forward, it looked nothing like the touring or sporting bikes that Harley-Davidson – and everyone else – was producing. If there was any doubt about who this bike was aimed at, the advertisements dispelled them: it referred to the Low Rider as 'The Mean Machine'.

Encouraged by the high sales achieved by this model, the company took bolder steps, and further designs by Willie G. appeared. The Fat Bob followed shortly afterwards, with its large dual tanks and high buckhorn handlebars, and in 1980 the no-holds-barred factory chopper, the Wide Glide, was introduced. With its wider and slightly longer front forks, stepped seat and backrest and the flames painted on the tank, there was no mistaking that Harley-Davidson was, at last, acknowledging just who was buying much of its output.

LEFT
The Albino Sturgis, created by John Reed, ex-patriot Englishman and parts' designer for Custom Chrome accessories. Apart from the tyres and exhausts, virtually everything is either gold-plated or painted white.

OPPOSITE LEFT
Goodman HDS-1200. For twice the price of a stock Sportster you will receive a bike that not only looks like an English cafe racer, but also handles like one. The Goodman company in England make replica frames for Manx Nortons and also aim to produce 100 of these Featherbed framed bikes every year.

Now that it was obvious where Harley-Davidson was coming from, the company showed just where it was aiming by giving the first Wide Glide off the line to *Easyriders* magazine for a road test; this was the first time that *Easyriders* had ever done a road test on a new Harley-Davidson. It did not take the Japanese manufacturers long to try to capitalize on the success that the new FXs were enjoying. Seeing that many motorcyclists wanted bikes that came straight out of the dealers shop looking like customized machines, they soon began producing their own imitations. They are still making such imitation bikes today, including a little 400c (25 cu in) Kawasaki, which is a direct copy of the 1340cc (80 cu in) Wide Glide but on a smaller scale, right down to the flames on the tank.

At one time a customized motorcycle was a byword for a chopper. Today, however, there are many different styles, and the only limitation on a bike is imposed by the owner's imagination. Many riders are content with the look of the stock

RIGHT
The 'Lowtail' that is built by OIT, the Harley-Davidson dealer in Breda, Holland. The bike is built around the special OIT designed 'Lowtail' frame and a 1340cc (80 cu in) Evolution engine. While every bike is built to individual requirements, most of them are similar to this one, with the 16-in (40-cm) front wheel and wide flat handlebars.

BELOW RIGHT
Tank art. Science fiction and fantasy scenes are perennially popular.

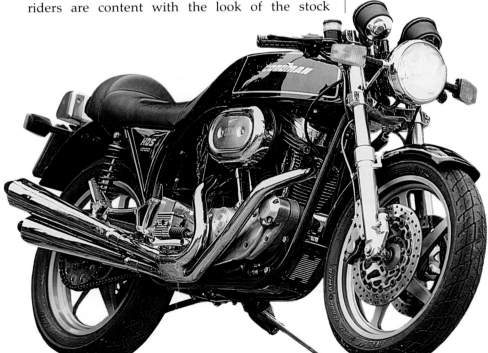

machines that Harley-Davidson produces today, and they are happy to make only slight cosmetic changes and performance improvements of the kind that can be purchased from the vast range of chromed extras and go-faster accessories that the company supplies.

For a significant number of owners, however, minor changes are not enough.

In the US, the style of customized machines is evolving continuously, and many additions or alterations are particular to certain parts of the country and are little seen elsewhere, which results in no one style being particularly typical. Moreover, fashions change so quickly that a custom house might bring out a new swing arm or set of exhausts or gas tank that will instantly become *the* essential item to have but, in a matter of months, will have become yesterday's news.

The general trend, however, is towards carrying out less extreme alterations to the forks and frames, while giving greater consideration to performance and handling and to small features and details.

It is not only in the US that styles of customizing vary from region to region. Some custom styles are especially popular in particular European countries. Sweden, for example, is renowned for turning out some of the best engineered and most radical customized machines around. It may be the long winter months, when there are few hours of daylight, and the short riding season that leave so much time for planning and building such detailed bikes, but there are certainly a great many for such a small country.

For many owners, possessing a Harley-Davidson means having a chopper that looks as clean and functional as possible. On the best examples this usually involves extending the Wide

Glide front forks by 20in (50cm), fitting a small, 80-spoke front wheel with no brake, a hardtail frame with no suspension, a solo seat and a minimal rear fender, and having the throttle as the only handlebar control.

In Germany there are strict laws which insist that any, even the most minor, modifications that are made to a bike have to be tested and approved. Riders, therefore, have to make many trips to a testing station during modifications, and this can demand a great deal of patience, dedication and expense on the part of the rider. For many German

and Dutch riders a large Shovelhead engine is the traditional choice. These are fitted into the lowest hardtail frame available, with the fattest tyres possible and every conceivable performance part added.

The French go more for style, and it is possible to buy a kit that will bolt on and instantly transform a Sportster into a mini-Fat Boy. French enthusiasts also build exact reproductions of bikes that look at first glance just like early Knuckleheads but that are, in fact, powered by the later Evolution engine.

BELOW
A Swedish Chopper. Long, low, lean and purely functional. A side stand is probably not necessary with a rear tyre of this width.

Many countries in eastern Europe – Poland and Czechoslovakia for example – had only old army WLs to work on. Many of these have been turned into small imitation Electra Glides by fitting them with replica fairings and panniers.

There will, of course, always be those though who regard any time spent cleaning and working on a bike as time wasted when they could be out riding instead, and they cultivate dirt and decay on their bikes. The best thing that can be said for this approach is that these Rat Bikes are the least likely to be the ones that are stolen.

EASYRIDERS

Easyriders magazine first appeared on the scene in 1971. It devoted itself exclusively to Harley-Davidsons and the custom world and has since grown to become the largest-selling motorcycle magazine in the world.

It has always championed the rights of the rider and tirelessly fought against unfair and sometimes dangerous government legislation, particularly the compulsory helmet law. (Harley-Davidson is, incidentally, the only motorcycle company to state that the rider should be able to decide whether to wear a crash helmet or not, although it does recommend that one is worn.)

Easyriders' relationship with the factory has not always been cordial. In July 1978 AMF discovered that an advertisement for Harley-Davidsons was to be placed in the magazine; the parent company ordered it to be withdrawn at the last moment, horrified lest its name be tainted by association with the magazine's readers. *Easyriders* left the empty page black but inserted a small warning in the centre of it stating that AMF considered the magazine hazardous to its image. After an outcry from Harley-Davidson dealers, who were only

too aware that this was the only publication that was promoting their motorcycles, a compromise was reached. Advertisements could appear in *Easyriders* as long as AMF was not mentioned and the advertisements were credited to the Harley-Davidson Dealer Network.

Easyriders has mellowed its editorial content over recent years, and it now reflects the many different types of people who ride Harley-Davidsons today. In consequence even the factory-sponsored Harley Owners Group advertises for members in its pages.

The magazine does deserve two special mentions. First, for recapturing the motorcycle land speed record from the Japanese in 1990 in its

Harley-Davidson powered 'streamliner'; and second, for the climax to its travelling motorcycle rodeos. In this, the final scene from the film *Easy Rider* is re-enacted: the two riders from the film roar into the arena on their choppers, trailed by the pick-up truck containing the gun-toting old men; the bikes are brought down, just as they are in the film, but this time they manage to shoot their assailants first, strap large American flags to their bikes and ride away as the crowd erupts. Revenge at last!

LEFT AND BELOW
A couple of bikes that show the different routes that builders can take to arrive at a similar destination. Both are built for straight-line performance. One for short trips up the drag strip, the other, with ten gallons of gas aboard, for those long highway runs.

ABOVE
Rat bike — it's under there somewhere. Surprisingly, none of the 'accessories' are available from the official parts catalogue! What these bikes have in their favour, though, is that they are the Harleys least likely to be stolen.

OPPOSITE
A proud owner with his Duo-Glide 'dressed' with 1400 lights.

What started out as an all-black XLCR Cafe Racer in 1977, has become something else entirely. All of the best customs succeed in provoking an extreme reaction and, while many will be appalled by such a bike, there will no doubt be those who will consider it a dream custom bike.

CHAPTER THREE

Racing

M otorcycle racing was taking place in the US before Harley-Davidson had even built its first machine in 1903. In sporadic, informal events a few riders would get together and charge around a horse-racing track or compete on an indoor bicycle velodrome, on what were little more than bicycles with small engines, at speeds that exceeded 30mph (48kph).

In 1909 motorcycle sport had an opportunity to become better organized when an English ex-cyclist, Jack Prince, opened the first motordrome in Los Angeles and soon afterwards rapidly constructed further 'dromes' in other cities throughout the US. These board tracks were constructed of 2 × 2in (5 × 5cm) or 2 × 4in (5 × 10cm) planks of wood, laid on edge and left rough for the tyres to grip. A circuit of these motordromes could be between a ¼ and 1½ miles (400m or 2.4km) in length, and they could be round or oval. The bends might be gently banked or nearly vertical. What they had in common, though, was the great danger to which they exposed the men who competed on them.

These thrilling races would attract crowds of up to 25,000 people, sometimes three times a week, to watch riders such as Cannonball Baker, Mile-a-Minute Collins and Dare Devil Derkum battle it out on Indians, Thors, Merkels and Popes. The riders' nicknames were not given lightly, as a crash on a board track was a frequent and often gory spectacle. A tumble on this rough timber, given the minimal protection that the riders wore, would often result in a body full of splinters – if the rider was lucky! Those who were less fortunate could be hit by one of the following bikes, whose rider's vision would be obscured by exhaust smoke and whose ability to manoeuvre on the oil-soaked track was severely restricted, because none of the bikes was allowed brakes.

When the crowds turned their backs on the carnage that was all too frequent at the motordromes in the 1920s, they turned to hill climbing for their thrills. While it was still spectacular, serious injuries were rare. The best hills were so long and steep that most competitors were unable to top them.

Joe Walter aboard a 1000cc (61 cu in) V-twin at the Speedway Park board track in 1915. Many of the races on these tracks were run over 300 miles (482.79km) and it was not unknown for a rider to make the two necessary pit stops for fuel and new tyres, pit again after suffering a puncture, survive a 100mph (160.93kph) fall and still get back on to win the race.

Glen 'Slivers' Boyd won his nickname after a fall on these boards, when he was dragged along underneath his bike. After a two-week stay in hospital, during which 200 splinters were removed from his body, he returned to the track proudly displaying the longest of the slivers that had been extracted – it was 14in (36cm) long! Not only was it dangerous for the riders, however. When Eddie Hasha lost control of his Indian at the Newark Motordrome in 1912, six spectators and fellow team-mate Johnnie Albright also died.

During these formative years of racing, when manufacturers were competing fiercely to win races with factory-sponsored teams, Harley-Davidson chose to remain apart. Although individual enthusiasts would enter, and often win, on a Harley-Davidson, they received no assistance from the factory. And then, much to the riders' annoyance, the company would boast in its advertisements that its machines were so good that they could win races without any support from the company.

In 1903 motorcycle clubs that had been meeting to coordinate races and to initiate safety regulations on a small scale founded the Federation of American Motorcyclists (FAM). It took several years for the FAM to be accepted by the majority of the riders, but by 1908 it was able to announce its first major event, a two-day endurance run around New York and Long Island. Harley-Davidson had been promoting its motorcycles as dignified and reliable transport rather than boasting of their sporting ability, and this was an event that could only enhance the company's claims.

President Walter Davidson, riding one of the early singles, lined up at the start as one of the 84 riders on 22 makes of machines. It was a tough course over unmade roads, and by the end of the first day many of the entrants had been forced to retire. After the completion of the 180-mile (290km) circuit of Long Island on day two, however, Davidson emerged the clear winner, with a perfect 1,000 point score. Encouraged by this victory, he remained in New York for another week to compete in an FAM economy run. In this he was first again, achieving an average of 188 miles to the gallon (68kmpl).

Noticing the increase in sales that followed these early competition successes, the company decided to enter the sport formally. To design racing motorcycles and set up a racing team, Harley-Davidson enticed William Ottoway away from Thor in 1913. Ottoway was one of the most intuitively gifted tuners of the time, but his engineering potential had not been realized at Thor, which was a small manufacturer. Working with William Harley, Ottoway began by modifying the 1000cc (61 cu in) pocket valve roadster, which was produced for the public. This tuned roadster was christened the Model II-K, and it was entered for two 300-mile (483km) races in the following year.

The first appearance was at Dodge City in Kansas, when Harley-Davidson unofficially supported six riders. This turned out to be a major disappointment – only two of the six bikes managed to complete the course. For the second outing at Savannah in Georgia later in 1914, the company employed some new riders and, for the first time, fielded an official team. This time it was more successful, with Irving Jahnke coming third. The rest of the team suffered from melting spark plugs.

Motorcycle sales had been increasing steadily over the years, and the market was dominated by the big three marques – Indian, Excelsior and Harley-Davidson. These companies now also started to dominate racing.

In 1915, for the first time, Harley-Davidson team riders appeared on the winner's rostrum when team captain Otto Walker and Red Parkhurst took the first two places in the 300-mile (483km) road race through the streets of Venice, California, on 28 March, averaging 63mph (101kph) around the 3-mile (4.8km) course.

July 1915 also saw one of the team's most celebrated victories at the Dodge City track. It was the same race around the 2-mile (3km) horse track at which Harley-Davidson had first made its disappointing debut the year before. This year's Independence Day competition attracted 15,000 spectators, who turned up to watch teams from Harley-Davidson, Indian, Excelsior, Cyclone, Emblem and Pope fight for the lead over 150 laps of the dusty circuit. Riding the 1000cc (61 cm in) pocket valve V-twins, Otto Walker crossed the finishing line first, with five of his team-mates in second, fourth, fifth, sixth and seventh places.

This Harley-Davidson team came to be known as the Wrecking Crew, because its collective appearance in a race would invariably 'wreck' the winning chances of most of the competitors. From 1916 the team increasingly dominated racing, until

in 1921 it won every single national championship – the 1-, 5-, 10, 25-, 50-, 100-, 200- and 300-mile titles, six of them at record speeds. In those days the championship was decided by a single race over each of the above distances run on the Syracuse dirt track in New York each year. Today, of course, becoming the dirt track national champion entails accumulating points from a series of races throughout the season.

Unfortunately, just as the Wrecking Crew hit its peak, sales of new bikes slumped, as we saw in Chapter 1, and at the end of 1921 the team was disbanded.

At this point in the story there is a natural pause, and it seems an ideal place to explain a little about flat track and the different classes of racing.

FLAT TRACK

Racing first began on horse tracks at county fairs at the turn of the century. On machines with rigid frames and no brakes, the riders would stay in top gear and pitch the bike sideways to slow down and negotiate the two 180-degree turns. (It was not, in fact, until 1969 that the American Motorcyclist

A smiling D H Davidson and fans in 1926. Five years previously, riding a Harley-Davidson on the same tarmac course, he became the first British rider to exceed 100mph (160.93kph). In America, the first officially timed motorcycle that reached this, was an Excelsior on a board track in 1912. (A V8-powered Curtiss had unofficially achieved 137mph (220.47kph) in 1907.)

Association (AMA) allowed the use of brakes on flat track bikes, which has changed the whole style of riding and the way in which the turns are negotiated.)

Variations on the basic flat track are the TT course, which is oval with at least one right-hand turn and one jump, and the short track, which is a ¼-mile (400m) circuit on which the smaller bikes, maximum 500cc (30.50 cu in) race.

Nowadays the AMA Dirt Track National Championship is fought out over 17 races, 15 of them over 1- and ½-mile (1.6-and 0.8-km) ovals, one short track and one TT course.

CLASS A

Only professional riders were eligible for Class A races, and they usually rode on specialized racing bikes. For safety reasons, when the large V-twins grew too powerful for small circuit racing, they were limited to 500cc (30.50 cu in) machines in 1924, and the limit was lowered to 350cc (21 cu in) in 1926. Hill climbers, however, were able to compete on 750cc (45 cu in) bikes.

After the demise of all the factory racing teams, interest in this class of professional racing declined until it was eventually terminated in 1938.

CLASS B

Class B races were for semi-professionals, and the bikes could be larger – 750cc (45 cu in) for circuits and 1340cc (80 cu in) for hill climbing.

Now that the Harley-Davidson racing team had been disbanded, Indian was the only manufacturer that was actively competing and consequently it enjoyed some of its most successful years in competition during the 1920s. Excelsior, the third largest manufacturer, had officially retired from

racing after its greatest rider, Bob Perry, died while testing a new 1000cc (61 cu in) model at the Ascot speedway in 1920. During this period Harley-Davidson also lost three of its best riders in accidents – Albert 'Shrimp' Burns in 1921, Ray Weishaar in 1924 and Eddie Brinck in 1927.

Each of these deaths generated a great deal of bad newspaper publicity, and during the 1920s there were increasing calls for something to be done about the number of fatalities and injuries that were occurring. The motordromes in particular were frequently referred to as 'Murderdromes', but their popularity was declining, and by the early 1930s only a few of the longer board tracks were left.

There was widespread agreement that the larger motorcycles were too powerful for the circuits on which they raced and that something had to be done to reduce the number of accidents. In response to this, in 1924 and 1926 the Motorcycle and Allied Trades Association, the organization that now controlled racing after the demise of the FAM, changed the rules governing Class A races. The new 350cc (21.35 cu in) series for Class A racing was formed after first Harley-Davidson and then Indian introduced overhead valve singles of this size to limit the speed on the small ½-mile (800m) tracks. This Peashooter, as the Harley-Davidson was nicknamed, dominated both short track and hill climbing when ridden by Joe Petrali. When not aboard this 350cc (21.35 cu in), Joe would ride the Excelsior Super X to more victories in the 750cc (45 cu in) events.

When Excelsior closed down in 1931 Petrali signed up with Harley-Davidson, and he went on to become the national champion in 1931, 1932, 1933, 1935 and 1936, in addition to the eight hill climbing championships he won between 1929 and

1937, riding first Excelsiors and later Harley-Davidsons. In 1935, at the annual national championship races held at Syracuse, New York, Petrali won the 1-, 5-, 10-, 15- and 25-mile races – all on the same day.

Joe Petrali was never really happy riding the larger Class C bikes, and when, while racing for the small bikes in Class A he was terminated in 1938, he gave up racing and eventually became a flight engineer for Howard Hughes.

CLASS C

This class of racing was introduced in 1934 in a bid to revive public interest and to encourage amateur riders back into the sport. Engine size was limited to 750cc (45 cu in) for side valves and 500cc (30.50 cu in) for overhead valve machines, with Novice, Amateur and Expert categories for riders.

The idea behind it was that only production motorcycles could be entered and that they had to be owned by the rider and ridden to the event they were competing in. Once at the venue, the machines would be stripped of their fenders, brakes and lights, and, still shod with their street-legal tyres, go straight out onto the track.

Well, that was the theory.

Class C racing rapidly became the most popular form of competition, for it brought racing back to its roots. Noticing the resurgence of interest after the demise of the professional Class A, both Harley-Davidson and Indian began to develop new 750cc (45 cu in) side valve V-twins. Indian delivered the Sport Scout and Harley-Davidson the WL, then, in 1941, the WR.

The WR was 'loosely' based on the uninspiring 750cc (45 cu in) D road bikes. I say 'loosely' because it was still the rule that the bikes should be stock production motorcycles, straight from the shop floor, and stripped of superfluous bits after the rider had ridden it to the race track. This rule was openly flouted, and although the bikes would look similar, internally they bore little resemblance to showroom stock. The WR came ready to race; nothing had to be removed.

Class C was quickly beginning to move towards becoming more like the recently deceased Class A professional racing.

With the WR, Harley-Davidson first used its particular style of designating the racing versions

of its production bikes. The W was the stock bike that the customer purchased; the WR was the dirt track model; and the WRTT was the version equipped for road racing. This method of identifying the variations of the same model has continued for all of its Class C racers – the K, KR and KRTT, the C, CR and CRTT, the XR, XRTT, and so on.

Shortly after the WR was introduced World War II intervened, and all competition was suspended. It was not until 1946 that a shortened racing season started up again, but by the following year everything was back to normal, and a full season was fought between the WR and the Indian Sport Scout.

THE AMERICAN MOTORCYCLIST ASSOCIATION (AMA)

Both Harley-Davidson and Indian could bend the rules that were supposed to govern racing because they ruled the AMA.

The Federation of American Motorcyclists (FAM) was formed on a very small basis in 1903 to promote the sport. When the FAM collapsed in 1919 because of lack of support from the riders, it was left to the Motorcycle and Allied Trades Association (M&ATA) to take over the promotion and running of the races, which it did reluctantly. M&ATA did not take on this burden of responsibility purely for philanthropic reasons, however, for, of course, it had a vested interest in the continued success of the sport.

A group of riders in Chicago wanted an organization that represented the men on the track as well, and so, in 1923, these riders created the American Motorcyclist Association to handle the competitive side of the sport. This arrangement lasted until 1928, when the riders again failed to support the AMA with their subscriptions.

LEFT
1968 KRTT ridden by Dan Haaby at Daytona that year. This was the first time that the Harley-Davidson colours of black and orange were used on the racing team bikes.

OPPOSITE
A 1961 KRTT. The seat hump on the rear mudguard was for the rider to slide on to, and enabled him to rest his chin on the tank for greater streamlining.

Someone had to come up with the money to keep the sport alive and organized, and it was left to the motorcycle manufacturers, or, rather, it was left to Harley-Davidson as the largest manufacturer, to finance it, with minor contributions being made by the financially troubled Indian and other small traders. So, as Harley-Davidson was paying the most to the AMA, it had a greater say when it came to the rules and how they were implemented. This dominance of the sport's governing body continued for many years, and it was only when some of the UK manufacturers were granted corporate membership of the AMA in 1961 that this set up started to change.

The K model made its racing debut in 1952, and for the first time featured a hand clutch, a foot shift on the right, swing arm and shock absorbers. Like its predecessor, the W model, it was also a 750cc (45 cu in) side valve, and, because the heads of the engine were squared off blocks of aluminium, it became known as the Flathead. Following the style set by the W series of bikes, the dirt track version was known as the KR, while the KRTT was for road and TT racing.

The all-new KR could produce only about 50bhp when it was first tested, scarcely more than the last of the WRs. But in the hands of the Harley-Davidson tuners it soon became a race winner. Suspension was still not used on dirt bikes, so when the rider brought his KRTT to the track to compete in these events he could remove the swing arm and shocks, and bolt on a rigid rear subframe in a matter of minutes. Then, after removing the fenders and brakes, the rider could be out on his practice lap. Although that is, clearly, something of an oversimplification of the preparations

that racers made, that was the idea behind class C racing. Anyone who wanted to compete need only buy one bike and, by substituting a few bits, could compete at dirt, pavement and TT events.

A third version, the KRM, was built for off-road racing. This was almost identical to the KRTT, but it featured different fenders, skid plate and a two-into-one exhaust. However, it proved to be far too heavy and unwieldy for competing in rough terrain and few were sold.

The KR could not arrive soon enough, for the 1952 season got off to a bad start at Daytona. All the Harley-Davidson riders were on the now outdated WRs, and the highest placing they had achieved was 17th, while Norton won for the fourth consecutive year. Towards the end of the year, however, the first KRs were entering and winning races, and early in 1953 Paul Goldsmith, riding a KRTT, won the Daytona race for Harley-Davidson for the first time in 13 years, with a record average speed of 94.45mph (153.58kph).

Sadly, this was also the year in which Billy Huber died from heat prostration while leading the 200-mile (322km) race in Dodge City. Billy was one of the most popular riders of the time, and he was, unwittingly, responsible for the small peanut gas tank on today's stock sportsters. This came about when he substituted the 4.5-gallon (20-litre) tank on his KR for the smaller, 2.5-gallon (11-litre) tank taken from a 125cc (7.6 cu in) Harley-Davidson two-stroke. It was adopted by the rest of the team, then by customizers, and it was eventually offered on the Sportster.

Before 1954 the national championships were decided by the annual races held on the mile (1.6km) track in Springfield, Illinois, irrespective of the rider's performance during the rest of the season. From 1954, however, it was on points

LEFT
The Aermacchi ERS TT 350cc.

OPPOSITE
A 1970 XR-750. This is what the 'iron head' XR looked like when it was first shown to the public. Although it had a single carburettor at its premier, by the time it was first raced there was one for each cylinder.

earned throughout the season at a variety of mile, ½-mile, road and TT events. In its first year it was won by Joe Leonard on a KR tuned by Tom Sifton. Sifton had sold his Harley-Davidson dealership in order to devote all of his time to tuning engines for the company and maintaining his reputation as the best in the business.

Joe Leonard won the championship in 1954 by coming first no fewer than eight times in 18 races. This record number of wins was eventually beaten in 1986 by Bubba Shobert, who scored nine wins, but from a series consisting of 33 races.

The following year this Leonard/Sifton KR engine was sold to the 19-year-old rookie, Brad Andres, who not only won the Daytona 200 with it but also the championship for 1955, with riders on Harley-Davidsons winning 15 of the 17 rounds. For

several years this winning streak continued unbroken. Leonard won again in 1956 and 1957; Carroll Resweber won in 1958, 1959, 1960 and 1961; and Bert Markel won in 1962. This golden era came to a close in 1962, however, when Resweber was forced to retire from the sport after receiving serious injuries in a pile up during practice at the mile track at Lincoln, Illinois, an accident that also claimed the life of Jack Goulsen.

At the same time, too, Japanese and English bikes were competing in, and increasingly winning, races. For many years Harley-Davidson had dominated domestic racing just as it had dominated the AMA and the rules controlling the sport, but by the late 1960s the tide was turning, as the AMA committee that controlled racing was restructured and chose its delegates from a broader base.

the Harley-Davidsons. This rule had been initiated in 1933 when Class C racing had begun and was long overdue for reform. For the 1969 season, therefore, Class C dirt track racing was open to any motorcycle of up to 750cc and in the following year the same rule also applied to road racing.

It was obviously time for the side valve KR to be replaced by a more modern overhead valve engine – no one seemed to have told Cal Rayborn, however, who rode his 'outdated' KRTT to victory at Daytona in 1968 and 1969.

Nevertheless, now that the British 750cc (45 cu in) bikes were allowed to compete, Harley-Davidson had to build a successor to the outdated K model. The need for a new racing bike could not have come at a worse time, for the company was in dire financial trouble and in the process of being taken over by AMF. The solution that was decided upon was to take the 883cc (55 cu in) XLR and, by destroking it, to reduce its capacity to 750cc (45 cu in), this was then slotted into the KRTT low boy frame and christened the XR-750. Equipped with an iron head (cast iron cylinders and barrels) and single carb, this was hardly more powerful than its predecessor, the KR.

The new XR-750 was first shown to the public in February 1970. Stylistically it was a great success; mechanically, however, it was going to be beset with problems. Although the XLR engine had performed well in the TT races for which it had been designed – that is, delivering short bursts of power – it just could not cope with being ridden flat out for mile after mile. Well aware that the XR was far short of being competitive, Dick O'Brien's workshop had been hard at work, for when the bike made its racing debut a few weeks later at Daytona in March 1970, it was already increased in power output and had a second carburettor fitted.

When the new committee met for the first time at the end of 1968 it abolished the rule that limited overhead valve bikes to 500cc (30.50 cu in) – that is, UK-made bikes – when side valve bikes were allowed a capacity of 750cc (45 cu in) – that is,

For the first time the British and the Japanese manufacturers could pit their 650cc (39 cu in) and 750cc (45 cu in) bikes against the Harley-Davidsons, and for the last time qualifying was determined by which bikes were the fastest. None of the Harley-Davidsons could qualify for any of the first 10 places on the grid. This was unsurprising as the XR could produce only 65bhp compared with the fastest qualifier, a Triumph 750cc (45 cu in) triple, which gave out 80bhp. During the race all four of the XR-750s overheated and broke down as they struggled to keep up with the pack. When Dick Mann on a Honda 750cc (45 cu in) took the chequered flag, an old KRTT ridden to sixth place by Walt Fulton was the best placing a Harley-Davidson achieved.

There were two reasons for these breakdowns. First, the mainshafts, which had been only pressed and not welded to the flywheels, came away, and this resulted in some spectacular internal damage. Second, the iron heads generated and retained far too much heat, and this caused the pistons to melt. The XR was christened the Waffle Iron by the unfortunate riders, who suffered repeated breakdowns throughout the rest of the year. At the end of the season, for the first time since the competition's inception, a Harley-Davidson did not occupy first or second place in the final championship placings. The despondent defending champion Mert Lawwill could only trail in in sixth position.

The 1971 season was equally disappointing. The bikes now had both carburettors on the right and one exhaust pipe running along each side and extra oil coolers fitted. But despite further internal modifications they still continued to break down and only ever registered one win.

Of the 200 iron head XR-750's that had to be produced to qualify it for racing, only about 100 were ever sold – the rest were scrapped.

In 1972 the rejuvenated version of the XR-750 arrived. After modifying the weaknesses inside the engine, the bore was enlarged and the stroke shortened to produce an over-square engine. Both carburettors were located on the right of the engine, and both exhausts now ran high along the left side. To crown the whole package, the iron barrels and heads were replaced by alloy versions with wider and more numerous cooling fins, hence the name Alloy Head.

This new version was immediately successful, and by the close of the season Harley-Davidson was back on the winner's rostrum, with Mark Brelsford taking first place and Lawwill, Rayborn and Sehl also appearing in the top 10.

During the early 1970s Japanese-made two-strokes swamped road racing, and threatened to take over the dirt track as well when Yamaha unleashed its four-cylinder TZ 750cc (45 cu in) two-stroke. This gave out a phenomenal 125bhp compared with the 88bhp that the XR now produced, and on the 1-mile (1.6-km) ovals it proved to be too dangerous to race on such a tight circuit. After Kenny Roberts wrestled the bike to its only dirt victory at the Indy Mile in 1975, the AMA brought in a new rule that only two-cylinder engines were eligible for dirt track races.

Throughout the rest of the 1970s Japanese bikes dominated road racing, while the XR-750 still had the edge on dirt tracks. Cal Rayborn, one of the greatest US riders of his generation, was tragically killed in 1973 while racing in New Zealand.

RIGHT
The XR 1000 was styled on the XR-750. This 1984 model has received many modifications including the red paintwork.

The 1986 season saw the championship divided into two separate series, road and dirt. The original idea of using the same bike in both types of event had long since proved to be impracticable, because the multi-cylinder two-strokes were the only contenders for road racing, and they were not allowed to compete on the dirt track, where the 750cc (45 cu in) V-twins ruled.

During the 1980s both Yamaha and Honda had introduced their own V-twins to contest the dirt races, but they enjoyed little initial success. However, Honda started to get things right, and between 1984 and 1987, it came out on top. The rest of the decade, though, was all Harley-Davidson's, and as the 1990 season ended and the XR-750 entered its 21st year of competition, Scott Parker was top man again for the third successive year – which is not bad for a 20-year-old engine up against the best that Japan can field.

One of the greatest growth areas in motorcycle sport is in the increasingly popular Sportster racing. The US Twin Sports Series began in 1989, and in its second year, such was the interest from riders and spectators, it could already boast a season of 34 races. It is exclusively for 883cc (55 cu in) Sportsters, which can be slightly modified, and it is run very much in the spirit of the original Class C rules. The engine is limited to a slight overbore, and most of the parts that are allowed to be replaced must come from the Harley-Davidson catalogue.

The great attraction of this type of racing is that all the machines start off equal, the winner is the person who is the best rider and who has the most innovative tuner. Sportster racing provides some of the closest and most exciting, not to mention the best sounding, racing around.

In the true spirit of Class C racing, that is just the way it should be.

Jay Springsteen, the National Champion in 1976, 1977 and 1978. The reigning champion rides with the No 1 plate while previous winners have one of the other single digit numbers, which they retain every year. Since his last championship win, Jay has suffered from a mysterious stomach ailment that invariably curtails his racing season. He has ridden with No 9 ever since his 1978 win and is probably still the most popular racer with the crowds.

The popularity of the 883cc Sportster racing is spreading throughout the world, and England, too, now has its own series. Pictured here is Grant Leonard who, when not racing, is a motorcycle journalist.

This is the Harley-Davidson-powered Streamliner that was put together by *Easyriders* magazine. It was built to snatch back the motorcycle land speed record from Kawasaki, and achieved this in 1990 by attaining 322.150mph (518.43kph), but not before a tyre blow-out caused the Stream-liner to crash at 300mph (482.79kph). Rider Dave Campos climbed out with only a bruised thumb to ask the video crew if they had captured it all on tape.

CHAPTER FOUR

Image

ABOVE
A scene from *Easy Rider*. When Captain America (Peter Fonda) and Billy (Dennis Hopper) are jailed for parading without a permit, they share a cell with a drunken lawyer, George Hanson (Jack Nicholson). Upon their release the following morning, he joins them on their trip to Mardi Gras.

George Hanson: 'Have *I* got a helmet!'

Once Harley-Davidson was just one of many motorcycle manufacturers in the US – as good as any and better than most. During the early years in which motorcycles were made, the machines appealed to the pioneer spirit that is inherent in the American people. Gas stations were non-existent, and a journey beyond the outskirts of town over rough tracks was often an adventure not to be undertaken lightly. Two wheels took over from four legs as riders swapped their horses for iron stallions and crossed the country over the old wagon trails to set new records.

The two runs that were the most hotly contested were the Transcontinental between California and New York, and the Three Flag Run, which was from the Canadian border to Tijuana in Mexico. The first of these transcontinental runs was in 1903, when George Wyman, riding a Duck motorcycle (later to be sold under the name of Yale), crossed the country in 53 days.

The prestige that went to the holder of these titles meant that no sooner had a Harley-Davidson, Excelsior or an Indian captured the record for the fastest run, than the other manufacturers would launch another rider to try to shave off an hour or two off the time, and by 1917 the record had been repeatedly broken and stood at 7 days and 16 hours.

When these riders were not criss-crossing the country, they were racing around the horse tracks or motordromes on bikes that had no brakes and minimal suspension. It took a great deal of courage and a fair amount of recklessness to compete in these events. Rivalry was intense and friendship was suspended when the starting flag dropped, as riders fought for the lead, often by bumping and elbowing their way through the pack and occasionally even trading punches at 100mph (160kph). Once off the track, these same men were a very close-knit group and only too ready to participate in a benefit race to raise funds for an injured rider's hospital bills or a widow's pension.

For many years the public was enthralled by reports of the latest cross-country endurance record, and people turned up in their thousands time and time again to watch a pack of riders charging around a race track. By the early 1920s, however, people had become increasingly sickened by the carnage on the tracks, and interest in the long-distance runs was declining. Road conditions were improving and cheaper cars and long-distance travel were a reality. Those who could afford transport were now choosing to travel on four wheels.

Motorcycles might easily have disappeared at this point. Indeed, manufacturers like Ace, Emblem and Reading Standard did close down, while the sales slump saw Harley-Davidson's output fall from an all-time high of 28,189 in 1920 to 10,202 in the following year. Motorcycles soon came to be widely regarded as acceptable for commuters, delivery boys or policemen, but anyone who rode a powerful bike by choice was often regarded at best as eccentric, and at worst as irresponsible and antisocial.

When Harley-Davidson was firmly established as a producer of quality and reliable motorcycles, the company began to emphasize the respectability and recreational aspects of motorcycling. Advertisements showed smartly dressed couples riding through the great outdoors or policemen patrolling the highways, guarding the nation against criminals and dangerous drivers. Those who rode machines with loud exhausts were called 'boobs' in these spreads and were requested to silence their machines. Women riders were encouraged, with claims that a Harley-Davidson was 'the woman's outdoor companion' and that it

'responds to the guiding hand of woman as did the kindest tempered steed of old'. The photographs that accompanied these advertisements would depict women, no longer confined to the sidecar or pillion seat, but capably handling their own big twin by themselves.

The Enthusiast, Harley-Davidson's own magazine, was first published in 1916, and it is still sent free to those who purchase new motorcycles and belong to official clubs. Published sporadically

LEFT
One of Florida's finest during the bike week that takes place at Daytona Beach every March.

RIGHT
The style of the covers of *Easyriders* magazine has changed somewhat over the years. Later editions see the bikes getting more of a look-in.

at first, it grew into a regular magazine that kept riders in touch with all that was happening in the world of Harley-Davidson. It was, and thankfully still is, a publication that cultivates and celebrates much of what is good about riding in the company of like-minded people.

Despite the advertising, after World War II the non-riding public's perception of Harley-Davidsons and motorcycling changed for the worse. There seems to be a correlation between soldiers returning from a war (World Wars I and II, Korea and Vietnam) and an increase in 'outlaw' motorcyclists' antisocial and illegal activities. There may be an academic thesis waiting to be written on the subject. Whatever the reasons behind them, however, some events, while insignificant in comparison with the behaviour of most motorcyclists, have had repercussions for everyone who rides, and particularly for those who ride Harley-Davidsons.

Legend has it that many of the 'outlaw' clubs that became infamous in later years, originated from the crews of bombers who had adopted the death-defying nicknames painted on the sides of their aeroplanes. When they were not fighting the war, they let off steam by riding around the west coast on motorcycles. After the war, many returning servicemen were demobbed in California with their severance pay, and they decided to settle there, buy a home and raise a family. Some of them though, unwilling to re-adjust to a 'normal' life, also decided to stay. They bought large motorcycles, usually Harley-Davidsons, and adopted more freewheeling lifestyles.

At weekends they would meet up and ride out to small towns in the country or cross the border into Mexico. Drinking and high spirits led to occasional confrontations with locals, and they would be run out of town by the local police. The result of all this would sometimes be a few arrests for drunkenness and brawling, a story in the local paper and townspeople tut-tutting about the youth of today, just as they have always done. Sometimes, however, they would turn up in large numbers at an AMA meeting and try to disrupt the races. It was at such a gathering that they first came to national prominence.

The events that occurred in the Californian town of Hollister in 1947 led to the first occasion on which one of these weekends received more than local attention. About 3,000 motorcyclists had turned up for the 4 July hill climb and track races, and by the evening many of them had congregated on the streets of the town. Drinking and racing in the streets led to confrontations with some of the local people, and fighting broke out. Just how many of these motorcyclists were actually involved in the trouble depends on whose account of the incident you believe. Some reports say that it took more than 400 local and state policemen to quell the disturbance; another, more dependable, source says that a force of 29 officers had the situation under control by the following morning. Whatever the truth about the extent of the troubles, a photographer from San Francisco who had arrived to cover the races for a local newspaper, went away instead with an unforgettable image that was reproduced on the cover of *Life* magazine, and on the inside pages a nation was informed about a potential 'threat' to society.

The film *The Wild One*, which was released in 1954, was based on these events, although it was banned in the UK until 1967. Viewed today it is hard to see what all the fuss was about. Once the motorcycles have faded into the background after

RIGHT
This picture, taken by a photographer at Hollister, appeared on the cover of *Life* magazine. This early 'bobber', stripped of superfluous parts, also has the handlebars mounted on risers. (© *San Francisco Chronicle*.)

So notorious did the film *The Wild One* become that it was banned in many places. To soften the film's image, some publicity material showed Marlon Brando riding in the company of women! Here he has even forsaken his leather jacket for clothing that could have been borrowed from his companions.

Lee Marvin reels from a punch thrown by Brando. While Brando rode a Triumph in the film, Lee Marvin rode a stripped-down Harley-Davidson.

Brando with his fellow gang members, The Black Rebels.

the first few minutes, it becomes just another small-town story of delinquent youths. Marlon Brando does give a towering performance of aimlessness, but he is totally devoid of any menace. Lee Marvin, on the other hand, looked and acted like the real thing, and he rode a Harley-Davidson.

While this very small yet troublesome element was creating headlines, many clubs across America carried on behaving as responsibly as they had always done. Many of them had their own strict dress code, and members wore uniforms that were almost military in appearance. Fines were even good-humouredly levelled against anyone who turned up having forgotten the club tie or hat. Weekends would see the club members congregating and riding off into the country, often to

attend a race meeting and to mix with other clubs or to participate in events to raise money for charity.

After the events at Hollister, however, police forces in California targeted the 'outlaw' clubs and made life as difficult as possible for them. Many of them disbanded, and the ones that continued kept a very low profile.

When two girls claimed they were raped at a Monterey, California, beach party held by one of these 'outlaw' clubs in 1964, state senator Fred Farr called for something to be done swiftly to combat this 'menace'. Although the charges were eventually dropped, the wheels had been set in motion, and the newly installed Attorney-General, Thomas C. Lynch, responded by sending out questionnaires to law enforcement officers all over the state. A few months later the replies were collated into a 15-page document which came to be known as the Lynch report.

One-third of this report was devoted to one particular club, and it gave a short, selective diary of events that had happened over many years, implying that this was just the tip of a huge iceberg,

and that 'outlaw' clubs were, in fact, responsible for many more crimes. When the report was picked up by the general media, its findings were distorted to give the impression that if the 'outlaws' were responsible for documented crimes, then ordinary riders were responsible for the rest! The salacious stories contained in the report made most of the newspapers in California when it was published, but, by the end of the week, it was largely forgotten – which is not surprising when you consider that over 200,000 motorcycles were registered in California at the time and that only a tiny proportion of these could be regarded as 'outlaws'. Motorcycling was generally regarded as a respectable form of recreation, indulged in by many.

And this is where the publicity should have ended. However, a newspaper correspondent sent a headline-grabbing story on the Lynch report to the *New York Times*, from where it was picked up by *Time* and *Newsweek* magazines, and millions of readers were given the impression that gangs of marauding motorcyclists were roaming about the country and that these roving 'outlaws' could turn up anywhere, at any time, and that your town might just be next. From here on politicians, newspaper editors and police chiefs manipulated such stories for their own ends. Some small newspapers fabricated reports out of nothing, proclaiming in front-cover stories that a holiday weekend had passed peacefully, without any motorcyclists arriving in town.

Hollywood, of course, saw a chance to make a quick profit and a whole slew of movies appeared in the late 1960s in the wake of this hysteria. These films were cheap to make and made a fortune: some local motorcyclists would be hired as extras and would be told to ride around a desert location while a banal story unfolded. Without ex-ception, the films presented an excess of sex, drugs, violence and bad acting, and it was all neatly justified at the end by showing the participants dying, jailed or sincerely repenting for any suffering they may have caused.

The first of these films, *The Wild Angels* (1966), was produced and directed by Roger Corman and involved many people who would later go on to bigger and better films, including actors Peter Fonda, Nancy Sinatra, Bruce Dern and Michael J. Pollard, assistant director Peter Bogdanovitch and editor Monte Hellman, who went on to make the seminal car road movie *Two Lane Blacktop* (1971).

Under the guise of moral outrage, *The Wild Angels* laid down the formula that other film studios later picked up, rewrote and filmed again in another desert location. What was probably the greatest cause of moral outrage to many though, was that the film was selected as the official US entry in the 1966 Venice film festival.

Hollywood had used Harley-Davidsons favourably in the 1930s and 1940s, usually showing policemen tearing down the highway in pursuit of the latest public enemy or as props in publicity photographs for actors and actresses, pictures of whom were often reproduced in *The Enthusiast*. Now, though, the image of a person who rode a Harley-Davidson was that of a rebel or undesirable, and those who believed the hype and were looking for something to rebel against, found a vehicle with which to express their antisocial inclinations.

When *Easy Rider* was made in 1969 it finally showed the Harley-Davidson customized motorcycles and the Harley-Davidson rider in a different light. For the first time Harley-Davidsons starred in a film that did not portray its riders as mindless thugs. In fact, the only thugs and hostile people in it were the 'normal' townsfolk whom Dennis

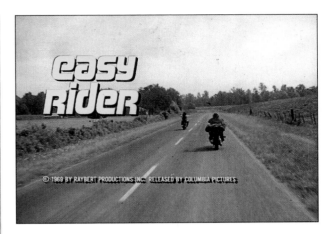

Hopper and Peter Fonda encounter on their ride to Mardi Gras. Most of these townsfolk were trapped and leading empty lives and resented the freedom that the two riders enjoyed.

The film came out at a time when many people were questioning their values, and many empathized with the sense of freedom shown by the two riders on the screen. Harley-Davidson picked up on this theme some time afterwards and began promoting its motorcycles as the 'great American freedom machine' and the Stars and Stripes as its main logo (remember the design on Peter Fonda's gas tank!). Now, the advertisements implied, even if you had a steady job and respon-sibilities, you could go out and experience this sense of freedom and still be back for work on Monday morning. Why, you could even experience it if you rode it home from the office on Monday evening. These were boom years for motorcycling, and Harley-Davidson sales almost tripled within four years.

There have been many films featuring Harley-Davidsons but most of them are only just watchable, if, that is, you close your eyes during the parts when the engines are turned off. Among them are *The Loveless* – a triumph of style over con-tent; *Girl on a Motorcycle* (1968) – Marianne Faithfull

RIGHT
Captain America and Billy share a reflective moment during an overnight stop on the road to New Orleans. George Hanson later articulates for them why they encounter so much hostility: 'They're not scared of you, they're scared of what you represent to them . . . what you represent to them, is freedom.'

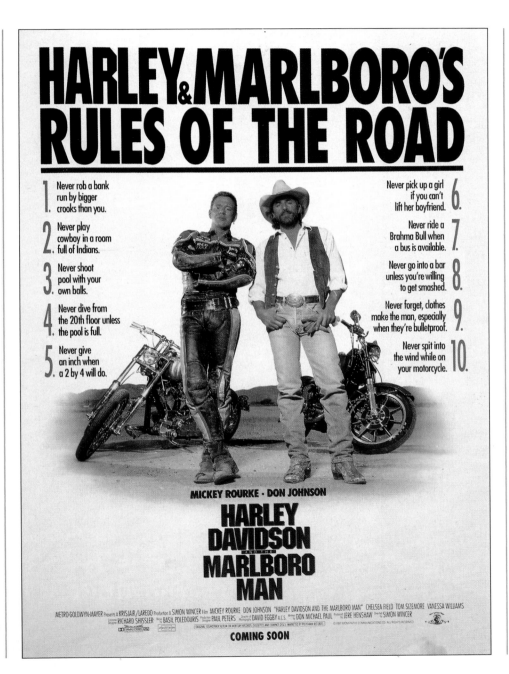

riding and writhing in her one-piece leather outfit; *Harley-Davidson and the Marlboro Man* – nice bike; *Electra Glide in Blue* (1973) – a film that is small, cute and unpretentious – rather like the hero; *Viva Knievel* (1977) – an ego in motion; *Code Two* – buddy cycle-cop saga; and *Streets of Fire* (1984) – *West Side Story* on wheels.

Of all the films that have starred a Harley-Davidson, only a few have presented the motor-cycle in anything like a favourable light. Notable amongst these was *Then Came Bronson*, which was first shown as a pilot film and then as a television series, running for 78 weekly episodes. Rather like a two-wheeled version of *Route 66*, *Then Came Bronson* starred Michael Parks riding around America on a Sportster doing good deeds and

LEFT
Mickey Rourke has ridden a Harley-Davidson in several of his films. 'It's a personal thing that can't be described. It is part of you. It's what you put into it, like your mind and your body.'

ABOVE
Scene from *Streets of Fire*.

Every Which Way But Loose. Clint Eastwood co-stars with an ape against a motorcycle gang. The picture says it all really.

Arnold Schwarzenegger in *Terminator 2* continues the current trend in films, for the good guy to ride a Harley-Davidson.

righting wrongs. As entertainment it was rather lightweight, but for the riding scenes alone it was compulsory viewing.

Peter Bogdanovich (remember *The Wild Angels*) directed Sam Elliot and Cher in the 1984 film *Mask*, which is one of the best screen depictions so far of Harley-Davidsons and their riders. While the characters in the film may not look or act like many who ride today, they are portrayed as the only decent and sympathetic people in an otherwise uncaring world.

Since the introduction of the Evolution-engined bikes in the 1980s the range of people who now own Harley-Davidsons has widened. Not only have many riders been attracted from foreign motorcycles to Harley-Davidsons, but there are also people returning to them for the first time in years, many of them attracted by the opportunity to own a new motorcycle that looks and sounds

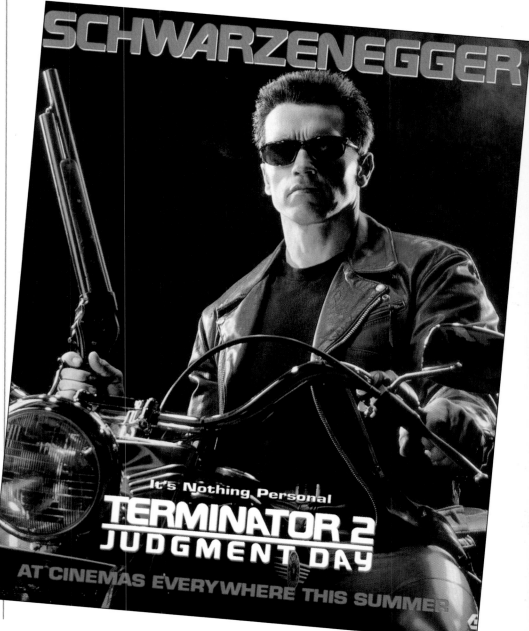

SCHWARZENEGGER

It's Nothing Personal

TERMINATOR 2
JUDGMENT DAY

AT CINEMAS EVERYWHERE THIS SUMMER

Stylish opening sequence from a television commercial for Peugeot cars. However, the line at the end of the film read 'Peugeot, when you are ready for four wheels'!

Reproduced from English *Vogue* magazine. Harley-Davidsons are today frequently used as props in fashion magazines and on record covers. Will the Harley-Davidson become just what the black leather jacket has turned into, a chic fashion item?

Gold halter-neck bodysuit, *both pages*, by James LoGerfo, £165, at Kumagai. Crystal earrings, by Eric Beamon, £65, at Harvey Nichols. Beauty note: all make-up by Elizabeth Arden, *including* Rose Petal Lipstick. Fake fur shawl-collared cropped jacket, *opposite*, by Liza Bruce, £240, at A La Mode. Gold leather sandals, *this page*, £185, at Manolo Blahnik.

Fashion editor: Lucinda Chambers. Hair: Syd Curry at Visages, LA. Make-up: Carol Shaw at Cloutier, LA

like a real motorcycle. For some people, having a Harley-Davidson parked in the garage next to an exotic sports car has become a symbol of achievement, showing that not only have they arrived but that they are also still young enough to indulge in that success.

This new type of owner has been reflected in advertisements and commercials the world over, which now seem to use Harley-Davidsons to help sell everything from chewing gum to cars. Not untypical is a television commercial for Levi jeans, which features a man riding a Harley-Davidson out of an elevator and cruising into a plush financial office. The rider throws a pair of Levis to a woman seated behind a desk, and she changes into them before climbing on to the pillion and riding away out of the office.

Scarcely a week passes without a spread appearing in a fashion magazine showing a model, wearing a ballgown or some other equally inappropriate attire, draped over a Harley-Davidson, or pictures of a new aftershave that does not use the traditional role models of a business executive or a muscular young man to sell the product, but instead, a man riding a Harley-Davidson down an empty highway.

This current interest in Harley-Davidsons and their use in advertisements has lasted for a few years now, yet it shows no sign of abating. What it proves though, is that the Harley-Davidson is no longer regarded as a vehicle for social outcasts. According to a survey carried out by Harley-Davidson, the average age of an owner is 34.4 years old, over half are married and earn an above-average income, while half again went to college or gained a degree.

CHAPTER FIVE

Riders

ABOVE
Unlike Daytona, where the wearing of club 'colours' is restricted, Sturgis is a lot more relaxed. The Vietnam Veterans are just one of the many clubs from all over the world that can be recognised on the streets. Membership of a Harley-Davidson club may start off as just a Sunday afternoon dalliance for some, yet it invariably becomes a whole way of life.

There is only one thing better than riding a Harley-Davidson and that is riding and socializing with other owners. While there are many loose-knit groups that meet occasionally, there are an increasing number of clubs that have been formed specifically to bring like-minded riders together. Some of these clubs were established many years ago, before the introduction of the Evolution engine and the many new riders that it has attracted.

Hundreds of thousands of Americans ride Harley-Davidsons, and there are clubs to cater for every interest – ex-Vietnam soldiers (The Vietnam Veterans), old bikes (The Antique Motorcycle Club of America), customs (The Modified Motorcyclists Association), motorcyclists rights (ABATE), to name but a few – as well as general clubs that welcome everyone.

Although Harley-Davidsons are popular all over the world, nowhere is there anywhere near the level of interest that exists in the US. Most other countries have national clubs with regional branches that accommodate everyone no matter what type of Harley-Davidson they ride. One of the oldest of these national clubs in Europe is the Harley-Davidson Riders Club of Great Britain, which has been organizing events for riders there ever since it was formed in 1948 and which is still thriving and attracting new members today.

The factory-sponsored Harley Owners Group was established in the US in 1983, and its membership has since topped 100,000. It is now beginning to expand into Europe after successfully establishing itself in the UK in 1990, where every Harley-Davidson dealer now has a HOG group. It held its first European rally there in 1991, and subsequently rallies will be held in a different European country every year.

ABOVE
In Great Britain, a year's membership of the factory sponsored Harley Owners Group (HOG) is a compulsory 'accessory' when a new bike is purchased.

OPPOSITE
For some the road goes on forever.

Countries as far afield as Poland, Czechoslovakia and Hungary, which at one time were virtually inaccessible to the touring rider, now host their own major annual rallies, in addition to the many smaller local events that are held throughout the year.

Touring around one of these countries in recent years, before the 'doors' were fully open, was quite an experience for both the rider and the people who lived there. The site of a large, modern Harley-Davidson riding through a town would literally bring the traffic to a halt as people stopped to stare at such a unique sight.

Many Harley-Davidson enthusiasts in Eastern Europe are still daily riding old WLs, left over from World War II, although the greater freedom they are now enjoying has meant that many more of them can now earn the money to purchase a more modern Evo. Many of these owners had, for many years, suffered constant harassment from the state and the police for just being different in their choice of transport or even for wearing a Harley-Davidson T-shirt.

STURGIS AND DAYTONA

Although the rallies now held in Europe are growing in size every year, with the largest ones now attracting around 4,000 people, they will never reach the scale of the two major American events, Sturgis and Daytona, which celebrated their 50th anniversaries in 1990 and 1991 respectively.

Sturgis Town, South Dakota, had grown up in the old gold mining days, but after the gold rush was over, the frontier moved on and it settled down to a quiet retirement, slumbering peacefully for many years. Some folk though were not content with such a quiet life and some of them were members of the local motorcycle club. It all began in 1936, when the Rapid City Motorcyclists, later renamed the Jackpine Gypsies Motorcycle Club, decided to renovate an old dirt track, which had been put out to grass many years before. Initially the club members held races for their own amusement but two years later the Black Hills Motor Classic was accepted as an official AMA ½-mile (200m) race for the 1938 season.

From humble beginnings, it grew from a weekend of racing and socializing for AMA clubs on a small scale, to a week-long party on a large scale with a little bit of racing here and there. Now Sturgis attracts Harley-Davidson riders from all over the world. It does draw riders of other makes of motorcycle as well, but if they turn up on anything that is not US-made, they tend to keep a low profile. If anyone were to turn up on a Japanese bike, they would be advised not to leave it parked on Main Street.

Sturgis hosts all of the usual attractions found at rallies – bike shows, bands and swap meets in addition to dirt and drag racing and some hill climbing action – but the overwhelming draw is to spend a week meeting old friends and making new ones, looking at the other bikes and riding around some stunning countryside, all with the beautiful sound of V-twin engines continuously reverberating in the background.

The Daytona Cycle Week is a year younger than Sturgis, but it evolved in a similar way – that is, around an annual AMA race, which was held on the beach. In 1960 the racing moved to the newly constructed speedway, 8 miles (about 12km) away, and the sandy beach was left for the visitors to profile up and down on, providing one of the few opportunities to ride in Florida without a crash helmet.

Harley-Davidson hosts a traditional show at the Hilton Hotel with the full range of new bikes on display. Out in the car parking lot there is a ride-in show for owners and test rides on the new models, all the proceeds going to the Muscular Dystrophy Association. Despite the presence of shows by Japanese manufacturers and a variety of different motorcycles on the street, Daytona remains predominantly associated with Harley-

Last Chance Saloon
in nearby Deadwood.

BELOW

Main Street, Sturgis.

Davidson, and it attracts riders from all over the US.

Away from the bike shows there are swap meets, drag racing on the beach and the chance to watch the non-stop parade of bikes cruising up and down Main Street for 24 hours a day. Oh, and there are the races at the speedway track, though for most of the visitors they are now little more than a small sideshow.

Daytona went through some difficult years in the early 1980s. The residents, mainly retired

BELOW

Anybody with the
right attitude fits in,
no matter who you
are or where you are
from.

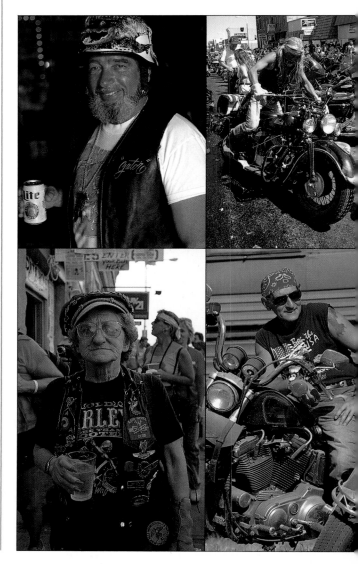

people, did not want the motorcyclists arriving for even one week a year. Policing became over-zealous and many riders began to refer to it as Daytona Cop Week. Fortunately, Paul Crow took over the Daytona Beach Police Department in 1988, and, after much liaison with all those concerned, he has done much to achieve a degree of tolerance by making the bikers feel welcome, the townsfolk unthreatened and by instructing the police to be firm but fair.

Sturgis, too, experienced problems for a while when the event grew to such a size that it became overwhelming for the local population. But, once again, there were people who were prepared to take the trouble to coordinate everyone's interests, and the results of their hard work and organization speak for themselves. For the 50th anniversary at Sturgis in 1990 between 300,000 and 400,000 motorcyclists were in town for a week-long party, yet at the end of the celebrations the only arrests were for traffic offences. As State Attorney General Roger Tellinghuisen told a *Rapid City Journal* reporter: 'the majority of people who attend the Motor Classic respect the law and extend courtesy to other rally goers. We have everything under control.'

LEFT
The annual ride out at Sturgis is one of the most popular events that is held. Each year the aim is for as many bikes as possible to congregate for a ride that often develops into a slow-moving traffic jam. Never have so many Harley-Davidsons got together and made so much noise or given their riders so much pleasure as they have done here.

NON-CIVILIAN SALES

The downturn in sales that occurred in the 1930s forced both Harley-Davidson and Indian to devote greater effort to selling to police forces. Although cars were increasingly favoured as transport for officers, motorcycles were ideal for highway patrol work and for catching speeders in the crackdown on highway fatalities, which had reached alarming proportions after many years of low-priority policing on roads that had no speed limits.

Harley-Davidsons were one of the fastest and most dependable vehicles on the road at the time and were, in fact, often used by bootleggers to transport their liquor in sidecar outfits. As the saying goes 'it takes one to catch one', and the Harley-Davidson was the favoured mount of many law enforcement agencies, with over 3,000 forces using them by 1940.

The competition for this market was intense, for not only did it account for many sales in these difficult years but the sight of policemen riding a Harley-Davidson also generated an enormous amount of good publicity. Few editions of *The Enthusiast* at this time would be published without a picture of yet another police force taking delivery of a new fleet. Prices were often slashed to capture this lucrative market, and occasionally, as a result of this cost cutting in the early days, policemen would be assigned to patrol the highways on very basic machines. 'Extras', such as a red light and siren, would often have to be bought out of the officer's own pocket.

LEFT
A member of the Los Angeles Police Department, ready to protect and serve his city.

Although Japanese manufacturers captured many of these sales in later years, Harley-Davidson has regained ground with the Evolution-engined bikes, and increasingly the Californian Highway Patrol is switching back to its traditional mounts.

Harley-Davidsons have, over the years, been used all over the world for law enforcement and ceremonial duties. The French police were riding them for traffic duties in Paris as early as 1919, while, at the time of writing (late 1991), Electra Glides are being tested for possible introduction into the UK.

Military needs have always required a motorcycle that can withstand the hardships of war and abuse by soldiers. Right from their first use by National Guardsmen during an abortive attempt to find Pancho Villa in northern Mexico in 1916, Harley-Davidsons have filled this niche for the American forces. They are said to have been imported by the Japanese army as early as 1912 and have since gone on to be purchased in large quantities for the armies in many countries including the UK, France, Australia, Russia and Canada, as well as for a variety of military units, from China to Mombasa.

Many men received their first experience of being aboard two wheels while they were in the army – during World War II alone, Harley-Davidson produced 90,000 motorcycles for the Allied forces. In addition to scouting and courier work, many were assigned to Cavalry Regiments to control tank movements.

Before they were let loose on the enemy, however, the soldiers were taught to ride on an eight-week course. Apart from learning how to stay on, they also had to learn how to fall off when the enemy was seen. One wonderful piece of contemporary newsreel shows a training officer firing a pistol into the air as a signal for a group of recruits to leap off their moving bikes and take cover while 30 speeding WLAs slide into the ground.

There have been many accounts, particularly in *The Enthusiast*, of soldiers' adventures while riding their Harley-Davidsons through difficult terrain for thousands of miles. Despite suffering appalling abuse and lack of servicing, the machines' ruggedness and dependability could always be relied upon.

BELOW
A soldier aboard his WLA 750cc. Note the ammo box on the fork leg and the black-out headlight.

The military front mudguard was wider and a greater distance from the wheel than the civilian version, so that it would not clog up in muddy conditions.

HARLEY-DAVIDSON TODAY

A visit to the factory leaves no doubt where the hearts of the workforce lie. A look around the parking lot is indication enough to the visitor where the hearts of the workforce are and of the pride they have in the product they make. A glance around the shopfloor will reveal that virtually every employee wears a T-shirt or belt buckle bearing a Harley-Davidson slogan. Gone are the AMF days when many riders would wear a T-shirt printed with the rather self-deprecating question 'If Harley-Davidson made aeroplanes, would you fly in one?' Workers on the factory floor today wear T-shirts that proudly boast 'Born in the USA' and 'God rides a Harley'.

This love of the Harley-Davidson does not stop at the boardroom door, either. For major rallies, such as Daytona and Sturgis, the business suits are swapped for leathers, as the directors saddle up and ride to the event. They also led the emotional ride from the York, Pennsylvania, plant to Milwaukee in 1981 to celebrate the buy-back from AMF, stopping off at every Harley-Davidson dealer on the way. Six years later they headed another mass parade, this time through the streets of New York and right up to the doors of the stock exchange on Wall Street when the company was

RIGHT
During bike week the whole town of Sturgis is given over to the visiting motorcyclists. Most of the town's shops are rented out to sellers of motorcycle products and the streets are filled with characters such as 'Fast Food Freddie'.

You do not usually have to seek out such people as they have a tendency of coming to find you.

approved for listing on the New York Stock Exchange.

The Muscular Dystrophy Association has been adopted by Harley-Davidson as its chosen charity, and over recent years it has organized many events that have contributed millions of dollars to the fight against the disease. The company's 85th birthday was celebrated by a mass cross-country ride to Milwaukee in 1988, with bikes appearing from every direction and every corner of the nation. When the contributions from the 40,000 riders who arrived were added up, over $500,000 had been collected. It is an on-going fund, and every HOG group in the world holds regular events to raise more money.

EMINENT ENTHUSIASTS

Apart from the many ordinary men and women who ride Harley-Davidsons – if, that is, there is anything 'ordinary' about such people – there are many others who are just as enthusiastic about their bikes and will ride them even when they have the option of travelling in a limousine.

One man who stood head and shoulders above the many celebrity enthusiasts and dilettantes was Malcolm Forbes, the owner of *Forbes* magazine, one of the richest men in the world and long-time Harley owner until his death in 1990 at the age of 70. Not only did Forbes have at least one Harley-Davidson stationed at each of his houses, but he also used to carry two with him on his yacht and one on his private jet, so that he might always have one around when the urge came on to go for a ride. As demanding as his business commitments were, when August arrived he would shelve all his business and social commitments and make the run to Sturgis with like-minded friends in the Capitalist Tools motorcycle club.

ABOVE
Malcolm Forbes also liked those around him to share in his love for Harley-Davidsons. When Elizabeth Taylor launched her perfume, Passion, he presented her with a Sportster painted in her favourite colour — purple — to match her product.

Forbes was not just passionate about Harley-Davidsons. He also had a collection of hot air balloons, one of which was a huge inflatable model of a Softail.

He was a great champion for the rights and safety of motorcyclists, turning up at conferences and using any means at his disposal to fight against unjust government legislation. When a group of senators submitted a resolution in 1980 calling for the establishment of a federal Strike Force to combat the threat of 'outlaw' motorcycle gangs, in defence of riders Forbes ran an editorial in his magazine stating that, just because some crimes are committed by people who ride motorcycles, it is not reason enough to harass the other seven million riders whenever a few of them get together for a ride. 'How about a strike force for law-breakers who also share an enthusiasm for horses or Chevrolets?' the editorial asked.

THIS PAGE AND OPPOSITE
Evil Knievel leaping over a long line of double-decker buses in England, mistiming his landing at the other end and sending the crowds home happy. He claims to have broken every single bone in his body at least once.

Evel Knievel enjoyed sponsorship from Harley-Davidson for many years when he rode an XR-750 to jump over an increasingly large number of vehicles. Even when he made his unsuccessful leap over the Snake River Canyon in 1974, the company logo was painted on the side of his jet-powered cycle. The partnership finally ended when Knievel was convicted of assaulting a business associate who co-authored a book that criticized the legend he had created.

Steve McQueen was a lifelong aficionado of US motorcycles, and he was planning to open a museum to show his large collection to the public

BELOW
Steve McQueen, on the right, aboard a police servicar on a hill high above San Francisco.

FROM OPPOSITE LEFT
Billy Idol, Mickey Rourke, Rob Halford of Judas Priest and Latoya Jackson. Is there anyone left in Hollywood who has not been pictured posing on a Harley-Davidson?

Former Hollywood actor Ronald Reagan made a presidential visit to the factory in 1987 to praise the workforce on their successful turn-around — a shining example of the quality and competitiveness of American industries.

RIGHT
Elvis Presley aboard
his 1956 KH.

BELOW
Get any group of
Harleys together and
not only will there
never be any two
identical bikes, but
you will also be
among some great
company. Here, the
'Sussex Coasters'.

before his death in 1981. He had over a hundred machines dating from many years awaiting renovation, in addition to the many examples that had been brought back to life for him by Sam Pierce, the former owner of the Indian Motocycle Company. After he gave up making films, he was frequently seen around Malibu riding one of his perfectly restored machines or just hanging out with the local bike crowd.

Even though it is now more popular than ever for a musician or film star to be photographed astride a Harley-Davidson to convey a certain image, there are many celebrities who do not climb off the bike once the camera is put away. Lou Reed, Bruce Springsteen, Dan Ackroyd, Sylvester Stallone, James Caan and Mickey Rourke are just a few of a huge number who ride Harley-Davidsons for pleasure, while Clark Gable, Roy Orbison and Elvis Presley are just a few of the many who are now riding elsewhere.

Index

ACKNOWLEDGEMENTS/ PICTURE CREDITS

p2 Martin Norris (bike courtesy Patrick Weeks); p3 Garry Stuart; p6–8 Garry Stuart; p9 (inset) The National Motor Museum at Beaulieu, (left) VEE Magazine; p10 The National Motor Museum at Beaulieu; p11 Grant Leonard; p12 l Pictorial Press, r A Morland; p13 EMAP Magazines; p14 The National Motor Museum at Beaulieu; p15 EMAP; p16 Grant Leonard; p17 VEE Magazine; p18 World's Motor Cycle News Agency; p19 Pictorial Press; p20 Andrew Morland; p21 bl VEE Magazine, t Martin Norris, br Pictorial Press; p23 The National Motor Museum at Beaulieu; p24 VEE Magazine; p25 EMAP; p26 t World's Motor Cycle News Agency, b Grant Leonard; p27 l Garry Stuart, r Garry Stuart; p28 l Martin Norris, r Garry Stuart; p29 Grant Leonard; p30 Grant Leonard; p31 VEE Magazine; p32–34 Martin Norris; p35 Andrew Morland; p36 Andrew Morland; p37 VEE Magazine; p38 VEE Magazine; p39 t Garry Stuart, b VEE Magazine; p40 Garry Stuart; p41 bl VEE Magazine, tr Martin Norris, br Pictorial Press; p42 VEE Magazine; p43 VEE Magazine; p44 Garry Stuart; p45 Andrew Morland; p46 Pictorial Press; p47 EMAP; p48 The National Motor Museum at Beaulieu; p49 The Hulton Deutsch Collection; p50 VEE Magazine; p51 The National Motor Museum at Beaulieu; p52 t VEE Magazine, b Garry Stuart; p53 t Garry Stuart, b VEE Magazine; p54–56 Grant Leonard; p57 The National Motor Museum at Beaulieu; p58/9 Grant Leonard; p60 World's Motor Cycle News Agency; p61 Grant Leonard; p62 VEE Magazine; p63 Joel Finlar; p64 Joel Finlar; p65 Garry Stuart; p66 l Garry Stuart, r Martin Norris; p67 San Francisco Chronicle; p68 Pictorial Press; p69 l Martin Norris, fl Joel Finlar; p63 Joel Finlar, t Pictorial Press; p71 Joel Finlar; p72 Pictorial Press; p73 l Martin Norris, b Joel Finlar; p74 al Joel Finlar; p75 Peugeot; p76 Martin Norris; p77 VEE Magazine; p78-83 Garry Stuart; p84 The National Motor Museum at Beaulieu; p85 The National Motor Museum at Beaulieu; p86 r Garry Stuart; p87 Pictorial Press; p88 The Hulton Deutsch Collection; p89 tl + b The Hulton Deutsch Collection; r Pictorial Press; p90-92 Pictorial Press; p93 t Pictorial Press, b Martin Norris.

Special thanks to VEE Magazine. All photos from VEE Magazine by Steve Berry and Robert Clarke.